Climate Change Management

Series editor

Walter Leal Filho, Hamburg, Germany

More information about this series at http://www.springer.com/series/8740

Jan Petzold

Social Capital, Resilience and Adaptation on Small Islands

Climate Change on the Isles of Scilly

Springer

Jan Petzold
Biosciences - Integrative Ecophysiology
Alfred Wegener Institute for Polar and Marine Research
Bremen
Germany

ISSN 1610-2010 ISSN 1610-2002 (electronic)
Climate Change Management
ISBN 978-3-319-52224-1 ISBN 978-3-319-52225-8 (eBook)
DOI 10.1007/978-3-319-52225-8

Library of Congress Control Number: 2017933433

Printed on acid-free paper

This Springer imprint is published by Springer Nature
The registered company is Springer International Publishing AG
The registered company address is: Gewerbestrasse 11, 6330 Cham, Switzerland

Acknowledgements

The three and a half years during which I worked on this book had naturally their ups and downs—with the former certainly overweighing. I had the privilege to conduct my research on the beautiful Isles of Scilly and temporarily also based in Cornwall. But apart from the natural environment, many people helped me to motivate, progress, rethink, present, discuss and, thus, finally accomplish my research.

First of all, I would like to thank Prof. Beate Ratter who introduced me to the world of island studies, advised me and constantly pushed my research forward.

I thank Prof. Catherine Leyshon and Prof. Stephan Harrison for the additional academic support and the opportunity to enhance my research with a stay at the Cornwall Campus of the University of Exeter. In Cornwall, I certainly want to thank also Kevin, as well as Sasa, for the great time in Falmouth.

My research would not have been possible at all without the help of the people of the Isles of Scilly. I benefitted from and enjoyed their welcoming attitude and openness. In particular, I want to thank Amanda Martin at the Museum of the Isles of Scilly for her time and support. Moreover, special thanks to Mark Prebble and Linda Owens, David Senior, Jonathan Smith, Clive Mumford and Richard McCarthy.

While the fieldwork on the Isles of Scilly and in Cornwall was the most exciting time of my research, most of my work was done in Hamburg, which I consider my home, and the University of Hamburg my home university. Therefore, I thank all the colleagues at the Institute of Geography as well as SICSS for the ongoing exchange of ideas and mutual support. Thanks to SICSS also for the financial support for additional fieldwork and conference trips.

Explicitly named in these acknowledgements should also be my patient proof-readers Corinna, Arnd, Matthias and Will. And, of course, Francesca—thank you for your help, patience, inspiration and love.

Contents

Abbreviations

AONB	Area of Outstanding Natural Beauty
AOSIS	Alliance of Small Island States
ATL	Advance the line
CBA	Community-based adaptation
COP	Conference of Parties
DEFRA	Department for Environment, Food & Rural Affairs
DTA	Duchy Tenants Association
EEZ	Exclusive economic zone
EMS	European Marine Site
EU	European Union
GCM	General circulation model
GDP	Gross domestic product
HEART	Honesty, Ethics, Accountability, Respect, and Truth/Transparency
HTL	Hold the line
IOSFGI	Isles of Scilly Farmers' and Growers' Initiative
IPCC	Intergovernmental Panel on Climate Change
ISREC	Isles of Scilly Renewable Energy Cooperative
LAG	Local action group
LDC	Least developed countries
MR	Managed realignment
NAI	No active intervention
NGO	Non-governmental organization
RSPB	Royal Society for the Protection of Birds
SES	Social-ecological system
SIDS	Small Island Developing States
SMP	Shoreline Management Plan
SPCK	Society for the Propagation of Christian Knowledge
SSSI	Site of Special Scientific Interest

List of Figures

List of Tables

Abstract

Small islands are among the regions most impacted by global climate change. Though regarded as particularly vulnerable, island societies exhibit a particular sociocultural resilience that distinguishes them from continental societies. This book addresses the questions of how small island communities deal with climate change, and how social capital helps to explain collective behaviour as well as opportunities and limitations of societal adaptability towards nonlinear environmental changes.

In the light of the social challenge presented by climate change, communities' responses become increasingly important. Besides formal and static adaptation planning, informal relationships affect communities' adaptability. Indeed, social capital can increase adaptability towards climate change. Yet how these relationships develop and their effects upon a community differ in every place and at every point in time. Equally varied are the impacts and features of environmental changes and those associated with climate change and sea-level rise. In addition, geographical context and changing environmental conditions influence the development of emergent social behaviour.

Complexity theory and resilience frame social capital in a more dynamic understanding by their connections to concepts of emergence, nonlinearity, path dependence and downward causation. Social capital theory from a complexity perspective, therefore, contributes to the understanding of micro-macro dynamics in social-ecological systems, as exemplified here for specified resilience towards sea-level rise on small islands.

In this book, I present the results of a case study on the Isles of Scilly, UK, which are among the most vulnerable islands in Europe. As is the case with the Isles of Scilly, many small island communities face environmental stressors, both historically and at present, which are expected to increase with further climate change. A mixed-methods qualitative and quantitative approach has been applied to analyse the various roles of social capital in this small archipelago's adaptation to climate change impacts, within its economically strong, yet seemingly peripheral and isolated context.

Chapter 1
Introduction

The years 2014 and 2015 were among the warmest since the beginning of systematic temperature recordkeeping in the nineteenth century. Sea surface temperatures and global sea levels were at a record high, tropical cyclones above average and sea ice at an extreme low (Blunden and Arndt 2015; World Meteorological Organization 2015). However, climate change has been a central environmental issue since the 1990s—in science, politics, and media. The topic is covered in various ways by almost every scientific branch. Its causes, developments and impacts, as well as related responses by politicians, media and populations provide a challenging basis for inter- and transdisciplinary research (cf. Klenk and Meehan 2015).

One of the central concerns regarding climate change is global sea-level rise and its impact on coastal areas and small islands. According to the Fifth Assessment Report by the Intergovernmental Panel on Climate Change (IPCC) it is 'virtually certain' that global mean sea-level rise will continue 'for centuries to millennia' (Stocker et al. 2013: 100) regardless of how efficiently mitigation targets are pursued. An observed global mean sea-level rise of 20 cm during the last century already resulted in coastal land loss on many islands (Forbes et al. 2013). Other impacts on coastal zones are the increased frequency and intensity of storm surges, increased salt water intrusion and the flooding of freshwater wetlands and valuable agricultural zones (Walker and Bellingham 2011: 298). A projected global mean sea-level rise of one metre or more until the end of the century will require effective adaptation in the most vulnerable areas—coastal zones and small islands (Nurse et al. 2014). Small islands have been identified as hotspots, due to their vulnerability, but also because they are good examples for the media and politicians to use to promote the agenda of environmental protection (cf. Kelman 2014).

Besides trying to prevent the worst impacts of a progressive climate change by mitigation efforts, global institutions and national governments integrate adaptation planning into their agendas, having recognised climate change as a major threat to future development that can no longer be impeded. Adaptation to climate change is associated with technological and socio-political measures (Nunn 2009).

J. Petzold, *Social Capital, Resilience and Adaptation on Small Islands*,
Climate Change Management, DOI 10.1007/978-3-319-52225-8_1

Large-scale sea defences, however, are hardly implementable in isolated and small island regions. Financial and physical resources are unavailable and the necessary infrastructure is unsustainable because large structural sea defences often cause further problems (e.g., erosion) which would have to be dealt with in the future (Kundzewicz 2002). It is a great challenge for island societies and decision-makers to find socially, ecologically and financially sustainable solutions to deal with pressures to their coastal habitats. Thus, effective planning and implementation in such a setting depend on the inclusion of community resources: local knowledge and skills and the integration of collective action for local adaptation, as well as the acceptance and support of communities for decision-making (Mercer et al. 2012; Jones and Clark 2013).

Discussions of environmental problems have long been centred upon loosely defined concepts such as vulnerability and sustainability. The discourse surrounding global climate change has largely followed this trend (cf. Adger 2006). Climate change research has primarily concentrated on understanding causes and modelling future scenarios. Impacts have always been on the agenda but have been associated with high degrees of uncertainty. In addition, the analysis of human and social dynamics in the context of climate change has long been neglected and the high relevance of the topic in media and politics has often been associated with a call for technical solutions and top-down adaptation programmes (cf. Berrang-Ford et al. 2015).

In response to these shortcomings, some researchers (e.g., Walker et al. 2004; Berkes et al. 2008; Glaser et al. 2008) have shifted their focus towards local contexts, regional dynamics and socio-ecological conditions. This has given rise to an alternative scientific perspective on adaptation to environmental change in terms of 'resilience', which stresses systems' behaviour and their capacities to adapt to changes. Such changes are considered to occur surprisingly and in a non-linear way, implying uncertainty. Resilience provides a framework for discussing change, dynamics and inter-connections in ecological and social systems. Analyses of the latter include a focus on the interaction between social agents and emergent collective behaviour, which shapes society (cf. Adger 2000; Berkes and Ross 2013). Nonetheless, resilience is a disputed concept, which has almost reached the status of the 'new sustainability'. It is often regarded as the solution to all environmental problems and it is associated with strengthening communities to tackle climate change (Weichselgartner and Kelman 2015). Do those communities need to be made more resilient—from the outside? Who defines what/who is or needs to be resilient?

Similarly, in the analysis of social relationships, the concept of social capital has to some extent developed into a mere buzzword, used in very different ways—but often superficially. Social capital has to be 'built' and 'promoted'. It delegates responsibilities to local communities. However, the dynamics and effects of social capital are highly dependent on socio-economic, cultural, historical and geographical contexts (cf. Naughton 2014: 5). Considering these contexts, in fact, offers a great potential in explaining and understanding social features and normative constructions of resilience.

In a world of change, coasts and small islands become interesting for researchers who discover the 'centrality of the marginal' (Gillis 2014: 164). Besides urban areas, rural and coastal regions are a focus of geographical studies. In this context, small islands are important objects of study due to their specific geographic factors and features of vulnerability (Moro et al. 2005), but are islands really particularly vulnerable or do they even all share a specific resilience? Approaches shaped by geographic deterministic presumptions, swayed by utopian and illusionist concepts of islands in literature and arts, neglect an expressive endeavour to critically study place-based communities and specific social-ecological conditions, as well as limitations and opportunities for adapting to environmental changes, such as global warming (cf. Petzold 2016). Thus, small islands must be studied 'on their own terms' (McCall 1994: 1) and in a non-deterministic way.

In this book, I focus on small island communities, collective action and adaptability towards sea-level rise. This implies understanding resilience in terms of capacity to cope with change. How does social capital help explain collective behaviour as well as opportunities and limitations of societal adaptability towards non-linear climate change on small islands?

The focus here lies not on institutions and planning but rather on social emergence and features of collective social capital—levels of trust, cooperation, collective action and social memory (cf. Young et al. 2014). Recent research on social capital and adaptation to climate change on islands has often focused on supposedly vulnerable Small Island Developing States (SIDS), e.g., in the Pacific (Hay 2003, 2013; Barnett and Campbell 2010; Kelman 2010). Yet few researchers have examined the role of social capital in dependent small island territories in the Western and European periphery (cf. Young et al. 2014). Rather than a quantitative cross-scale comparative study in this work I focus on an archipelago, allowing for research of islands on their own terms as well as overcoming a constructed island-continent and island-sea dichotomy (McCall 1994; Stratford et al. 2011).

My research is based on empirical fieldwork conducted on the Isles of Scilly (Fig. 1.1), a group of five inhabited small islands, which are highly exposed to storm surges and vulnerable to sea-level rise—hazards that are expected to increase with future climate change. Due to their geological history and their vulnerability to further sea-level rise, the islands have been described as a 'drowned landscape' (Thomas 1985) and the 'Maldives of the Atlantic' (Anon 2009). On the one hand, the islands depend upon external assistance, especially in terms of financial resources for large-scale problems. On the other hand, they depend upon their own capacity to solve problems (cf. Grydehøj 2013). The Isles of Scilly share typical features of small dependent island territories worldwide that are vulnerable to the projected impacts of global climate change, albeit, in contrast to the discussion on SIDS (cf. Hay 2013), in a comparatively financially rich context. Using qualitative and quantitative research methods, I analyse the potential and limitations of various dimensions of social capital with regard to climate change adaptation on the Isles of Scilly (cf. Petzold 2016).

My research contributes to the further refinement of social capital theory and social emergence. Using a complexity theory approach allows me to apply the full

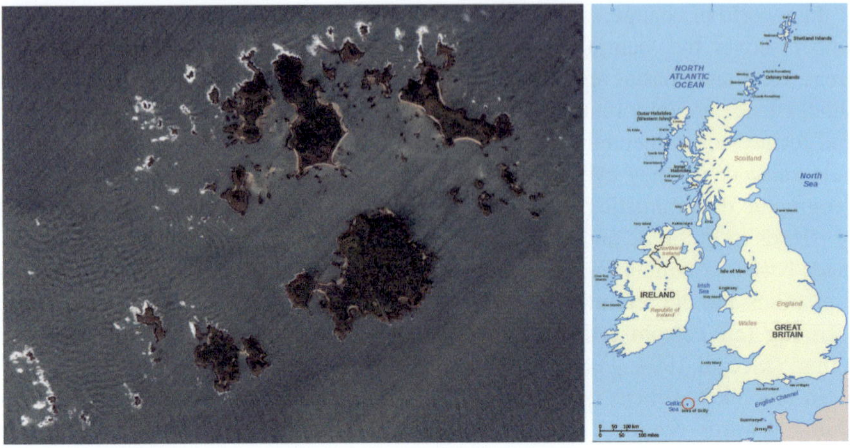

Fig. 1.1 Satellite image and location map of the Isles of Scilly (*Source left*, NASA 2007; *right*, Ruhrfisch 2010)

analytical value of the multi-dimensional concept of social capital to a case study. In this way, I intend to contribute to filling the theoretical gap between individual interactions and an emerging social structure (cf. Sawyer 2005).

Combined with the perspective of resilience, I show how social capital can help to explain how societies develop beneficial behaviour for climate change adaptation. This approach is also relevant from a policy standpoint, since the social sciences are still underrepresented in climate change research (cf. Victor 2015) and the recent agenda of the international community not only increases efforts at climate change mitigation but also provides more assistance to vulnerable regions.

The perspective of island studies with its associated concept of nissology contributes a critical conception of geographic places and system boundaries. Nevertheless, the relatively young fields of island studies and nissology need further empirical and theoretical foundations for the concepts of 'islandness' and archipelagos.

This book is composed of seven chapters. Chapter 2 provides a brief overview of the latest insights from climate change research, including expected impacts and implications of uncertainty, with a focus on sea-level rise. Chapter 3 integrates three different streams of theory, beginning with the theoretical concepts of social-ecological resilience, complexity and social emergence, and a focus on coastal communities. This lays the basis for a comprehensive review of social capital theory, including the identification of research gaps and potentials. Furthermore, this chapter discusses small islands, how they can be understood as special systems, and how 'place matters' in geographical research. The chapter concludes by uniting the three streams of theory and directing them toward the Isles of Scilly. As a result, specific research questions arise which are in turn posed to the island communities themselves. Chapter 4 establishes a detailed framework for answering these questions, including my epistemological approach, justifications for the choice of the case study, and a set of qualitative and quantitative strategic research methods from the social sciences. The core part of this work is Chap. 5. Here, I present the full range of empirical data collected. The chapter includes a detailed account of the isles' environmental and societal background, levels of social capital and potentials for adapting to climate change. This allows for the specific research questions from Chap. 3 to be answered. Chapter 6 returns to the theoretical approaches. It abstracts from the case study and explains what we could learn from the empirical results, how the different theoretical approaches contribute and interact and how the results help to advance theoretical development. This will finally shape the conclusions in Chap. 7, including outlooks for further research opportunities and the relevance of the findings for communities and policy-making with regard to climate change adaptation, both on small islands and in general.

References

Adger, W. N. (2000). Social and ecological resilience: Are they related? *Progress in Human Geography, 24*(3), 347–364.

Adger, W. N. (2006). Vulnerability. *Global environmental change, 16*(3), 268–281.

Anon. (2009). *Scilly Isles could have to be abandoned because of global warming*. The Telegraph. http://www.telegraph.co.uk/news/uknews/5024580/Scilly-Isles-could-have-to-be-abandoned-because-of-global-warming.html, checked on November 05, 2015.

Barnett, J., & Campbell, J. (2010). *Climate change and small island states: Power, knowledge, and the South Pacific*. London, Washington: Earthscan.

Berkes, F., Colding, J., & Folke, C. (2008). *Navigating social-ecological systems. Building resilience for complexity and change*. Cambridge: Cambridge University Press.

Berkes, F., & Ross, H. (2013). Community resilience: Toward an integrated approach. *Society & Natural Resources, 26*(1), 5–20.

Berrang-Ford, L., Pearce, T., & Ford, J. D. (2015). Systematic review approaches for climate change adaptation research. *Regional Environmental Change, 15*(5), 755–769.

Blunden, J., & Arndt, D. S. (2015). State of the climate in 2014. *Bulletin of the American Meteorological Society, 96*(7), ES1–ES32.

Forbes, D. L., James, T. S., Sutherland, M., & Nichols, S. E. (2013). Physical basis of coastal adaptation on tropical small islands. *Sustainability Science, 8*(3), 327–344.

Gillis, J. R. (2014). Not continents in miniature: Islands as ecotones. *Island Studies Journal, 9*(1), 155–166.

Glaser, M., Krause, G., Ratter, B. M. W., & Welp, M. (2008). Human/nature interaction in the anthropocene. Potential of social-ecological systems analysis. *GAIA, 17*(1), 77–80.

Grydehøj, A. (2013). One archipelago, many approaches. Diverse systems of governance and economy in the Isles of Scilly, UK. In *International conference on Island development*. October 15, 2013. Penghu, Taiwan: IGU Commission on Islands.

Hay, J. E. (2003). *Climate variability and change and sea-level rise in the Pacific Islands region. A resource book for policy and decision makers, educators, and other stakeholders*. Tokio, Japan: SPREP.

Hay, J. E. (2013). Small island developing states: Coastal systems, global change and sustainability. *Sustainability Science, 8*(3), 309–326.

Jones, N., & Clark, J. (2013). Social capital and climate change mitigation in coastal areas: A review of current debates and identification of future research directions. *Ocean and Coastal Management, 80,* 12–19.

Kelman, I. (2010). Hearing local voices from small Island developing states for climate change. *Local Environment, 15*(7), 605–619.

Kelman, I. (2014). No change from climate change: vulnerability and small island developing states. *The Geographical Journal, 180*(2), 120–129.

Klenk, N., & Meehan, K. (2015). Climate change and transdisciplinary science. Problematizing the integration imperative. *Environmental Science & Policy, 54,* 160–167.

Kundzewicz, Z. W. (2002). Non-structural flood protection and sustainability. *Water International, 27*(1), 3–13.

McCall, G. (1994). Nissology: A proposal for consideration. *Journal of the Pacific Society, 17*(2–3), 1–14.

Mercer, J., Kelman, I., Alfthan, B., & Kurvits, T. (2012). Ecosystem-based adaptation to climate change in Caribbean small Island developing states: Integrating local and external knowledge. *Sustainability, 4*(12), 1908–1932.

Moro, D., Agard, J., & Vicente, L. (2005). Island systems. In R. M. Hassan, R. Scholes, & N. Ash (Eds.), *Ecosystems and human well-being. Current state and trends, Volume 1: Findings of the condition and trends working group of the millennium ecosystem assessment* (Vol. 1, pp. 663–680). The millennium ecosystem assessment series. Washington, DC: Island Press.

NASA. (2007). ISS014-E-16597. In *International space station imagery*. http://spaceflight.nasa.gov/gallery/images/station/crew-14/html/iss014e16597.html, checked on December 18, 2015.

Naughton, L. (2014). Geographical narratives of social capital: Telling different stories about the socio-economy with context, space, place, power and agency. *Progress in Human Geography, 38*(1), 3–21.

Nunn, P. D. (2009). Responding to the challenges of climate change in the Pacific Islands: Management and technological imperatives. *Climate Research, 40,* 211–231.

Nurse, L., McLean, R. F., Agard, J., Briguglio, L., Duvat, V., Pelesikoti, N., et al. (2014). Small islands. In V. R. Barros, C. B. Field, D. J. Dokken, M. D. Mastrandrea, K. J. Mach, T. E. Bilir, et al. (Eds.), *Climate change 2014: Impacts, adaptation, and vulnerability. Part B: Regional aspects. Contribution of working group II to the fifth assessment report of the intergovernmental panel on climate change* (pp. 1–60). Cambridge, New York: Cambridge University Press.

Petzold, J. (2016). Limitations and opportunities of social capital for adaptation to climate change: A case study on the Isles of Scilly. *The Geographical Journal, 182*(2), 123–134.

Ruhrfisch. (2010). Isles of Scilly locator map. Available under creative commons attribution-share alike 3.0 Unported license: https://creativecommons.org/licenses/by-sa/3.0/deed.en. Wikimedia Commons. https://commons.wikimedia.org/wiki/File:Scilly_Isles_Locator_Map.png#, checked on October 19, 2016.

Sawyer, R. K. (2005). *Social emergence. Societies as complex systems* (Vol. 4). Cambridge: Cambridge University Press.

Stocker, T. F., Qin, D., Plattner, G.-K., Alexander, L. V., Allen, S. K., Bindoff, N. L., et al. (2013). Technical summary. In T. F. Stocker, D. Qin, G.-K. Plattner, M. Tignor, S. K. Allen, J. Boschung, et al. (Eds.), *Climate change 2013: The physical science basis. Contribution of working group I to the fifth assessment report of the intergovernmental panel on climate change* (pp. 33–115). Cambridge, UK, New York: Cambridge University Press.

Stratford, E., Baldacchino, G., McMahon, E., Farbotko, C., & Harwood, A. (2011). Envisioning the archipelago. *Island Studies Journal, 6*(2), 113–130.

Thomas, C. (1985). *Exploration of a drowned landscape. Archaeology and history of the Isles of Scilly*. B.T. Batsford: London.

Victor, D. G. (2015). Climate change: Embed the social sciences in climate policy. *Nature, 520* (7545), 27–29.

Walker, L. R., & Bellingham, P. (2011). *Island environments in a changing world*. Cambridge: Cambridge University Press.

Walker, B., Holling, C. S., Carpenter, S. R., & Kinzig, A. (2004). Resilience, adaptability and transformability in social–ecological systems. *Ecology and Society, 9*(2), 5.

Weichselgartner, J., & Kelman, I. (2015). Geographies of resilience: Challenges and opportunities of a descriptive concept. *Progress in Human Geography, 39*(3), 249–267.

World Meteorological Organization. (2015). *2015 likely to be Warmest on record, 2011–2015 Warmest five year period.* https://www.wmo.int/media/content/wmo-2015-likely-be-warmest-record-2011-2015-warmest-five-year-period, checked on December 18, 2015.

Young, E., Muir, D., Dawson, A., & Dawson, S. (2014). Community driven coastal management: An example of the implementation of a coastal defence bund on South Uist, Scottish Outer Hebrides. *Ocean and Coastal Management, 94,* 30–37.

Chapter 2
Climate Change: Impacts, Uncertainties and Adaptation

Climate scientists are becoming increasingly certain of the drivers and impacts of global climate change, while models for projecting future climate change are progressively improving. Nevertheless, significant uncertainties inevitably remain and there is a lack of robust data on the regional level. Small islands, on the one hand, are often represented as drowning lands and climate change's first victims due to sea-level rise. On the other hand, detailed data about the impacts of climate change are hardly available for small islands. Individually they have little power to deal with long-term adaptation by themselves.

The findings of climate science provide sufficient evidence to show that mitigation measures alone—efforts to keep emissions levels and thus global warming below a certain level—are not enough. Attentions should be turned towards adaptation—adaptation in circumstances of uncertainty. Adaptation to climate change is not only a question of political strategies and technological solutions, but it is also a social challenge. It is important to look at the regional and local level of communities and to understand potentials and obstacles of societal action as well as the necessity and circumstances of adaptation measures. This is where social sciences and integrative, interdisciplinary research play a major role.

Therefore, in addition to an overview of current insights in climate sciences, I also provide an overview of sea-level rise, not least of all because it will probably be the 'most immediate, most certain, the most widespread, and the most economically visible' (Pilkey and Young 2009: 4) impact of climate change. In addition, a focused account of the implications of the 'greatest potential threat' (Byrne and Inniss 2002: 10) to small islands will be given in Sect. 3.3.1.

J. Petzold, *Social Capital, Resilience and Adaptation on Small Islands*,
Climate Change Management, DOI 10.1007/978-3-319-52225-8_2

2.1 Observations and Projections

According to the IPCC '[h]uman influence on the climate system is clear' (IPCC 2013: 1); the causes of changes in mean air temperatures are 'extremely likely' (IPCC 2013: 5) to be linked with anthropogenic influences on the composition of the earth's atmosphere since the beginning of the industrial era. Greenhouse gas emissions can be attributed mainly to countries from the so-called global north (Victor et al. 2014: 113). Mean surface temperatures, widely viewed as an indicator of climate change, show an increase of 0.85 °C between 1880 and 2012 (Stocker et al. 2013: 37). Warming oceans and melting glaciers and ice sheets have already caused a rise in the global mean sea level of 0.19 m between 1901 and 2010 (Stocker et al. 2013: 46). Further observations of climate change impacts include ocean acidification and warming of oceanic water masses, leading to coral bleaching and changing patterns of fish stocks. Similarly, changing atmospheric temperatures and precipitation patterns influence the composition of flora and fauna, causing shifts in climatic zones. An increase of extreme weather events, such as droughts and tropical cyclones, is often linked with climate change, too. Observations show such increases, but the causes are still not completely established and trends differ greatly by region (Stocker et al. 2013: 50).

Based on so-called general circulation models (GCM), climate scientists develop projections for possible futures under climate change. Such models help us to better understand the physics of the climate system and deduce future developments. The earth's climate system is understood as a complex system. As such, climate change models can only have limited predicting power due to unknown thresholds, abrupt changes and features of chaos. Accepting these limitations of climate models 'does not negate their value in informing society about possible futures. However, it does suggest that these models may not be preparing us for some possible responses' (Harrison and Stainforth 2009: 111). The improvement of climate models has also helped to define the existing uncertainties. The question is how we deal with uncertainties. To allow for effective political action, it must be clear that '[w]e do not need to demand impossible levels of certainty from the models to envisage a better safer future' (Maslin 2013: 270).

The main impacts expected of a further rise in global mean surface temperatures (Fig. 2.1) are increased extreme temperatures, further ocean acidification and a rise in global mean sea level of almost one metre (depending on the model) until the end of the twenty-first century (IPCC 2013: 10–13).

2.2 Sea-Level Rise

It 'is *virtually certain*' (Stocker et al. 2013: 100, emphasis in original) that sea-level rise will continue 'for centuries to millennia' due to the inertia of the climate system and the ongoing thermal expansion of the oceans, irrespective of mitigation targets

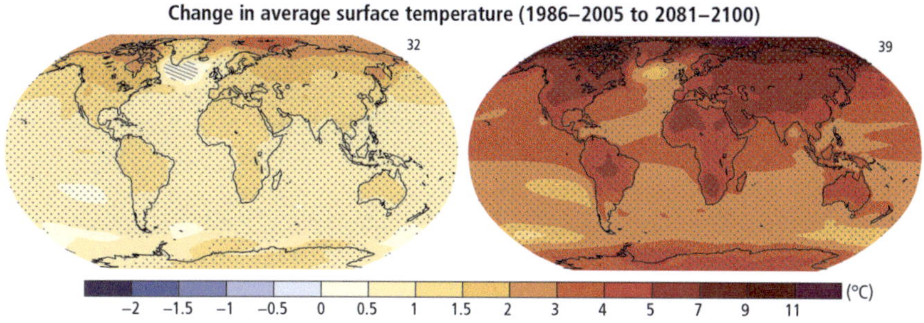

Fig. 2.1 Projected change in average surface temperatures between 1986–2005 and 2081–2100 (*left* RCP2.6, *right* RCP8.5) (*Source* IPCC 2014: 61)

set by the international community. Mitigation of greenhouse gas emissions is nonetheless important in preventing further long-term climate change. A warming of mean surface temperature by 2–4 °C compared to pre-industrial levels might even cause thresholds to be surpassed, threatening irreversible changes to global circulations and climate systems, because '[t]he long-tailed uncertainty implies that there is a considerable risk that relative sea level rise will exceed recent high-end scenarios' (Grinsted et al. 2015: 21). Such thresholds could imply a complete melting of the Greenland or West Antarctic ice shields and consequently cause a sea-level rise of up to 3–7 m (cf. Schaeffer et al. 2012, Stocker et al. 2013: 100).

To understand the implications of sea-level rise, it is important to distinguish between eustatic and isostatic sea level changes. The former results from a changing volume of water masses or accommodation space in oceans due to tectonic plate movement. The latter is caused by relative sea level falls due to rising land as the result of glacial melting (Pilkey and Young 2009: 23). Factors such as surface winds, freshwater influx, water temperature and ocean currents influence relative sea levels. Thus, changes in sea level are not evenly distributed across the planet: 'regional sea level changes may differ substantially from a global average' (Church et al. 2013: 1191). Moreover, sea-level rise due to thermal expansion is a linear event, in contrast to sea level-rise resulting from melting ice sheets (Pilkey and Young 2009: 65). Projections show that sea-level rise can be expected in most regions across the oceans, except regions near glaciers and ice sheets, where a sea-level fall is projected. The highest regional deviations above the global mean are projected to be in the South Pacific and South Atlantic, as well as around North America. Near the Arctic and Antarctica, projected sea level changes lie as little as fifty per cent below the global mean (Fig. 2.2).

In addition to the uncertainties and inaccuracies of modelling and regional downscaling of sea-level rise projections, the actual impact of sea level changes on coasts differs from place to place. Vulnerability to sea-level rise does not only depend on the magnitude of climate-induced increases in sea level. Geomorphologic and non-climate coastal drivers influence how coastlines adapt to changes in sea level. The impacts of sea-level rise will especially affect

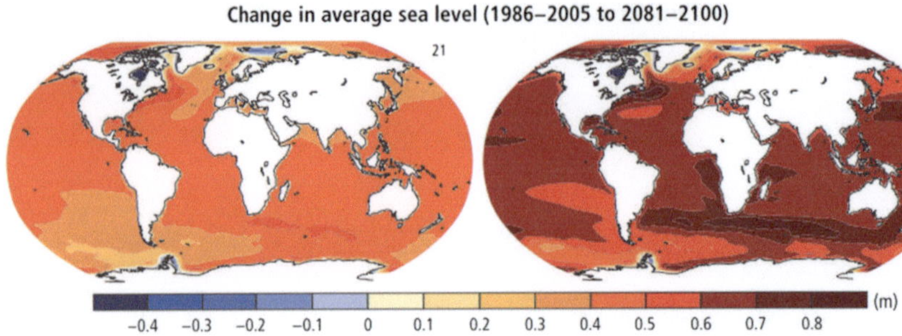

Fig. 2.2 Projected change in average sea level between 1986–2005 and 2081–2100 (*left* RCP2.6, *right* RCP8.5) (*Source* IPCC 2014: 61)

low-elevation coastal zones, such as deltas, which are often also densely populated (e.g., in South, Southeast, and East Asia). Moreover, small and low-lying islands, such as those found in the Caribbean, and Indian and Pacific Oceans, are among the most affected regions (Nicholls and Cazenave 2010: 1519).

2.3 Adaptation

Apart from climatic changes and other natural pressures on coastal ecosystems, however, a decisive factor on how coasts adapt to sea-level rise is human action—something that is not included in climate models. Many problems associated with sea-level rise are in reality problems caused by human action, such as removal of coastal vegetation, sand mining or maladaptation involving seawalls. Mangroves are used as materials for the construction of houses and boats; wetlands are cleared for aquacultures; beaches are built up for tourism infrastructure (cf. Pilkey and Young 2009: 108). On the other hand, the vulnerability of coastal zones also depends on non-climate and non-natural factors, such as population density and existing adaptive capacity, which includes many factors (see below). Seen from this perspective, small islands are the most vulnerable regions to sea-level rise (Barnett and Adger 2003). A major problem of the IPCC projections for coastal management is that they focus on changing mean sea levels rather than peak tides and major flood events, which are most relevant to coastal communities (Hinkel et al. 2015: 188).

Despite uncertainties, and because of regional differences regarding the impacts of climate change, adaptation planning is the main task of climate scientists, next to physically understanding and modelling climatic changes, since '[a]n improved understanding of adaptation is fundamental, because it is one of the biggest determinants of actual rather than potential impacts' (Nicholls and Cazenave 2010: 1519). The IPCC defines adaptation as '[t]he process of adjustment to actual or expected climate and its effects' (Noble et al. 2014: 838). However, the 'diversity of local conditions is one crucial barrier to adaptation' (Eisenack 2012: 108). Therefore, effective adaptation relies on the assessment of adaptation needs, which in turn requires information on risks and vulnerability for the biosphere, society, institutions and the private sector, as well as identifying technological and financial capacities. This assessment is multi-dimensional and multi-scalar, considering the regional, national and international levels, and is the basis for the development and choice of adequate adaptation pathways (Noble et al. 2014: 839–844).

Two major categories of adaptation to climate change impacts are technological/structural and management/socio-political measures (Nunn 2009; Noble et al. 2014). Nunn (2009) identifies technological/structural measures such as the restoration of ecosystems (e.g., mangrove forests), changing agricultural crops in response to temperature variability, infrastructure improvement, protection and relocation in response to increased storm surges, natural conservation of ecosystems, relocation of settlement and coastal protection in response to sea-level rise. Management/socio-political imperatives, on the other hand, include national strategy plans and long-term planning, hazard mapping, agricultural management, sponsorship of research, legislation regarding construction and marine protected areas (Nunn 2009: 222; cf. Petzold and Ratter 2015).

The range of adaptation options, however, is much wider than the few examples given above. The purpose of assessing the adaptation needs of a certain place is to identify effective solutions. The variation in responses to sea-level rise exemplifies how certain adaptation options that are realistic in one region will not work in others (e.g., the Netherlands, unlike Bangladesh, has hundreds of years of experience with dyking, technical knowledge, financial resources and necessary infrastructure to implement solutions, including dykes, building codes, warning systems and shelters against sea-level rise). In poor countries, responses to sea-level rise are often more effective since the more radical solution (i.e., usually location) is chosen, whereas rich countries tend to prefer costly but short-term solutions, such as raising seawalls and dykes (Pilkey and Young 2009: 168).

Moreover, adaptation needs must be identified so that maladaptation can be avoided. According to the IPCC 'maladaptation refers to actions, or inaction that may lead to increased risk of adverse climate-related outcomes, increased vulnerability to climate change, or diminished welfare, now or in the future' (Noble et al. 2014: 857). Examples of maladaptation are the construction of hard infrastructures, such as seawalls, while neglecting negative long-term impacts, whether on the ecosystem or on social structures. Maladaptation can also be caused by taking action too late or too early due to lack of information, or by applying traditional solutions that are no longer sufficient (Noble et al. 2014: 858). Protective measures against sea-level rise, such as seawalls or artificial beaches created by sand nourishment, also generate a sense of security and encourage construction, which can have negative impacts that were unforeseen when the adaptation measure was planned (Pilkey and Young 2009: 166). Copying successful adaptation measures from one region to another can also lead to unnecessary environmental damages if they are not adjusted to local circumstances (Pilkey and Young 2009: 135–136).

2.4 Community-Based Adaptation

According to Adger et al. (2013: 112), after National Research Council (2010) research and policy on adaptation and mitigation have problematically 'largely focused on the material aspects of climate change, including risks to lives and livelihoods, the costs of decarbonising economies and the costs of impacts on various sectors of the economy'. Adaptation, however, is always culturally and socially framed. Within this context, I concentrate on social features, which can constrain but also enhance adaptation options. As the large body of vulnerability literature shows, social contexts influence perception of risks, preference of adaptation option and actors involved (Blaikie 1994; Cutter 1996; Adger 2006). Social structures, distribution of poverty and education also influence the degree of vulnerability and adaptive capacity of different members of a community. Moreover, sense of place affects how people accept certain adaptation options, including migration (Klein et al. 2014: 915–916). Certain features of social structures can also be advantageous in promoting adaptation measures, including distribution of local expertise, willingness within the community to commit to collective action, and high levels of trust and reciprocal help (cf. Petzold and Ratter 2015). Because of the inherent limitations of sea-level rise projections, vulnerable communities must live with a degree of uncertainty and accept expert qualitative estimates (Pilkey and Young 2009: 58; Barnett and O'Neill 2010). Decision-makers and planners can benefit from integrating community resources into socially, ecologically and financially sustainable solutions for dealing with changes to coastal habitats. Effective solutions translate local knowledge and collective action into local adaptation measures, while raising acceptance and supporting communities in the implementation of adaptation measures (cf. Mercer et al. 2012; Jones and Clark 2013).

Community-based adaptation (CBA) is an approach that combines adaptation and development while benefiting from local resources and ensuring effective implementation and outcomes of adaptation measures. CBA 'addresses the locally and contextually specified nature of climate change vulnerability' and stresses the role of 'participatory processes, involving local stakeholders and development and disaster risk reduction practitioners, rather than being restricted to impacts-based scientific inputs alone' (Ayers and Forsyth 2009: 26). Such community-based approaches are often developed by local or international non-governmental organisations (NGO) in cooperation with local communities by holding consultation meetings, conducting household surveys and developing tools and technologies for adaptation measures that can be implemented by the community itself. Examples are floating gardens and the construction of elevated houses in flood-prone areas, or the establishment of sharing networks (e.g., for food, tools, and skills) in isolated communities (Ayers and Forsyth 2009; Dodman and Mitlin 2013). Difficulties associated with CBA can be problems of governance and relations of power that constrain such bottom-up initiatives (Dodman and Mitlin 2013). Moreover, there is often a lack of data regarding future impacts of climate change, which makes it difficult to distinguish between community-based development and adaptation. It is also difficult to monitor and learn from, or 'upscale', CBA projects due to their place-specific character. Last but not least, these projects might not suffice for long-term climate change if emission levels continue to rise (Ayers and Forsyth 2009: 28–29).

Social sciences contribute to the analysis of potentials and limitations of adaptation options, by identifying driving factors and obstacles in a regional context. This can help in implementing adaptation measures in an effective and socially sound way. Social science methods can produce essential insights that complement the data gained from climate change models, projections and quantified vulnerability assessments.

References

Adger, W. N. (2006). Vulnerability. *Global Environmental Change, 16*(3), 268–281.

Adger, W. N., Barnett, J., Brown, K., Marshall, N., & O'Brien, K. (2013). Cultural dimensions of climate change impacts and adaptation. *Nature Climate Change, 3,* 112–117.

Ayers, J., & Forsyth, T. (2009). Community-based adaptation to climate change. *Environment: Science and Policy for Sustainable Development, 51*(4), 22–31.

Barnett, J., & Adger, W. N. (2003). Climate dangers and atoll countries. *Climatic Change, 61*(3), 321–337.

Barnett, J., & O'Neill, S. (2010). Maladaptation. *Global Environmental Change, 20*(2), 211–213.

Blaikie, P. M. (1994). *At risk. Natural hazards, people's vulnerability, and disasters.* London: Routledge.

Byrne, J., & Inniss, V. (2002). Island sustainability and sustainable development in the context of climate change. In H.-H. M. Hsiao, C.-H. Liu, & H.-M. Tsai (Eds.), *Sustainable development for Island societies. Taiwan and the world* (pp. 3–30). Taipei: Asia-Pacific Research Program.

Church, J. A., Clark, P. U., Cazenave, A., Gregory, J. M., Jevrejeva, S., Levermann, A., et al. (2013). Sea level change. In T. F. Stocker, D. Qin, G.-K. Plattner, M. Tignor, S. K. Allen, J. Boschung, et al. (Eds.), *Climate change 2013: The physical science basis. Contribution of working group I to the fifth assessment report of the intergovernmental panel on climate change* (pp. 1137–1216). Cambridge, UK, New York: Cambridge University Press.

Cutter, S. L. (1996). Vulnerability to environmental hazards. *Progress in Human Geography, 20*(1), 529–539.

Dodman, D., & Mitlin, D. (2013). Challenges for community-based adaptation: Discovering the potential for transformation. *Journal of International Development, 25*(5), 640–659.

Eisenack, K. (2012). Archetypes of adaptation to climate change. In M. Glaser, G. Krause, B. M. W. Ratter, & M. Welp (Eds.), *Human-nature interactions in the anthropocene. Potentials of social-ecological systems analysis* (Vol. 1, pp. 107–122), Routledge studies in environment, culture, and society New York: Routledge.

Grinsted, A., Jevrejeva, S., Riva, R., & Dahl-Jensen, D. (2015). Sea level rise projections for northern Europe under RCP8.5. *Climate Research, 64*(1), 15–23.

Harrison, S., & Stainforth, D. (2009). Predicting climate change: Lessons from reductionism, emergence, and the past. *Eos, Transactions American Geophysical Union, 90*(13), 111–112.

Hinkel, J., Jaeger, C., Nicholls, R. J., Lowe, J., Renn, O., & Peijun, S. (2015). Sea-level rise scenarios and coastal risk management. *Nature Climate Change, 5*(3), 188–190.

IPCC. (2013). Summary for policymakers. In T. F. Stocker, D. Qin, G.-K. Plattner, M. Tignor, S. K. Allen, J. Boschung, et al. (Eds.), *Climate change 2013: The physical science basis.*

Contribution of working group I to the fifth assessment report of the intergovernmental panel on climate change (pp. 3–29). Cambridge, UK, New York: Cambridge University Press.

IPCC. (2014). *Climate change 2014: Synthesis report. Contribution of working groups I, II and III to the fifth assessment report of the intergovernmental panel on climate change*. Geneva. http://ipcc.ch/pdf/assessment-report/ar5/syr/AR5_SYR_FINAL_All_Topics.pdf, checked on September 23, 2015.

Jones, N., & Clark, J. (2013). Social capital and climate change mitigation in coastal areas: A review of current debates and identification of future research directions. *Ocean and Coastal Management, 80,* 12–19.

Klein, R. J. T., Midgley, G. F., Preston, B. L., Alam, M., Berkhout, F. G. H., Dow, K., et al. (2014). Adaptation opportunities, constraints, and limits. In C. B. Field, V. R. Barros, D. J. Dokken, K. J. Mach, M. D. Mastrandrea, T. E. Bilir, et al. (Eds.), *Climate change 2014: Impacts, adaptation, and vulnerability. Part A: Global and sectoral aspects. Contribution of working group II to the fifth assessment report of the intergovernmental panel on climate change* (pp. 899–943). Cambridge, United Kingdom, New York, NY, USA: Cambridge University Press.

Maslin, M. (2013). Cascading uncertainty in climate change models and its implications for policy. *The Geographical Journal, 179*(3), 264–271.

Mercer, J., Kelman, I., Alfthan, B., & Kurvits, T. (2012). Ecosystem-based adaptation to climate change in Caribbean small Island developing states: Integrating local and external knowledge. *Sustainability, 4*(12), 1908–1932.

National Research Council. (2010). *America's climate choices: Adapting to the impacts of climate change*. http://www.nap.edu/catalog/12781/americas-climate-choices, checked on March 05, 2015.

Nicholls, R. J., & Cazenave, A. (2010). Sea-level rise and its impact on coastal zones. *Science, 328* (5985), 1517–1520.

Noble, I. R., Huq, S., Anokhin, Y. A., Carmin, J., Goudou, D., Lansigan, F. P., et al. (2014). Adaptation needs and options. In C. B. Field, V. R. Barros, D. J. Dokken, K. J. Mach, M. D. Mastrandrea, T. E. Bilir, et al. (Eds.), *Climate change 2014: Impacts, adaptation, and vulnerability. Part A: Global and sectoral aspects. contribution of working group II to the fifth assessment report of the intergovernmental panel on climate change* (pp. 833–868). Cambridge, United Kingdom, New York, NY, USA: Cambridge University Press.

Nunn, P. D. (2009). Responding to the challenges of climate change in the Pacific Islands: Management and technological imperatives. *Climate Research, 40,* 211–231.

Petzold, J., & Ratter, B. M. W. (2015). Climate change adaptation under a social capital approach —An analytical framework for small islands. *Ocean and Coastal Management, 112,* 36–43.

Pilkey, O. H., & Young, R. (2009). *The rising sea*. Washington, DC: Island Press, Shearwater Books.

Schaeffer, M., Hare, W., Rahmstorf, S., & Vermeer, M. (2012). Long-term sea-level rise implied by 1.5 °C and 2 °C warming levels. *Nature Climate Change, 2*(12), 867–870.

Stocker, T. F., Qin, D., Plattner, G.-K., Alexander, L. V., Allen, S. K., Bindoff, N. L., et al. (2013). Technical summary. In T. F. Stocker, D. Qin, G.-K. Plattner, M. Tignor, S. K. Allen, J. Boschung, et al. (Eds.), *Climate change 2013: The physical science basis. Contribution of working group I to the fifth assessment report of the intergovernmental panel on climate change* (pp. 33–115). Cambridge, UK, New York: Cambridge University Press.

Victor, D. G., Ahmed, E., Dadhich, P. K., Olivier, J., Rogner, H.-H., Sheikho, L., et al. (2014). Introductory chapter. In O. Edenhofer, R. Pichs-Madruga, Y. Sokona, E. Farahani, S. Kadner, K. Seyboth, et al. (Eds.), *Climate change 2014: Mitigation of climate change. Contribution of working group III to the fifth assessment report of the intergovernmental panel on climate change* (pp. 111–150). Cambridge, UK, New York: Cambridge University Press.

Chapter 3
Social Capital and Small-Island Resilience

If island populations are, in fact, faced by serious impacts of climate change, and the social and cultural challenge of adaptation is recognised, how can we understand how these changes influence our societies? How do societal actions influence our natural environment? What can we learn from past interactions between populations and nature? I already pointed out the relevance of regional differences and my focus on a special geographical land type—small islands. How can one define and distinguish *small* islands and how do we have to study them?

This chapter provides the theoretical framing needed to answer these questions, a basis for the development of adequate research methods and interpretation of data. To analyse both the potential and the limitations of social factors with regard to adapting to pressures associated with climate change on small islands—especially sea-level rise—this chapter introduces theoretical approaches from various perspectives.

The first section concentrates on the questions of how communities adapt to changes in social-ecological systems. The concept of emergence can work as an analytic lens for understanding societal action, positive and negative feedbacks, and implications of human-environment interactions for resilience.

Secondly, I integrate the concept of social capital, to better understand limiting and promoting factors that influence social emergence and communities' capacities to adapt to change. Social capital offers the perspective on social structures and networks, which, in turn, must be analysed within their geographical, political, and cultural contexts.

Thirdly, I analyse the role of island-specific perspectives on vulnerability and resilience, and the critical construction of small islands as isolated and peripheral regions. Nissology provides a critical perspective on islands as closed and vulnerable systems, which is necessary for analysing and abstracting from specific island case studies.

The fourth section brings together these three sections of critical in-depth literature review and definitions of key concepts as a basis for the development of

J. Petzold, *Social Capital, Resilience and Adaptation on Small Islands*,
Climate Change Management, DOI 10.1007/978-3-319-52225-8_3

research questions and adequate research methods and for the operationalisation that will then be applied to the case study.

3.1 Social-Ecological Resilience

As laid out in the previous chapter, climate change and climate change adaptation imply a high degree of uncertainty with regard to forms and geographical distribution of impacts, combined with regionally different socio-economic and institutional responses (cf. Eisenack 2012: 107). Since rigid planning might not help to deal with uncertain future developments, adaptation from a resilience perspective focuses on coping with change. This section deals with the question of how societies in so-called social-ecological systems cope with change.

Focussing on adaptation to environmental changes, like climate change, involves the concepts of vulnerability and resilience. These concepts are closely related but take different perspectives. Adger (2006: 269) defines vulnerability as 'the degree to which a system is susceptible to and is unable to cope with adverse effects (of climate change)'. Central parameters for vulnerability, in general, are a system's exposure to stress, sensitivity and coping capacity. The concept of resilience specifically addresses the latter. It is a concept with origins in ecology (Holling 1973) and in developmental psychology and mental health research (Masten 2001). The former perspective provides the basis for the concepts for adaptability in social-ecological systems and is deeply rooted in complexity theory. Here, it refers to 'the shocks and stresses experienced by the social-ecological system, the response of the system, and the capacity for adaptive action' (Adger 2006: 269).

Therefore, the key ideas of complex systems will be laid out first here, with a focus on social emergence. Social emergence points out how social relationships and interactions produce social structures, which cannot be explained by individual attributes and behaviour. Subsequently, I introduce the concepts of resilience, adaptability, and transformability—concepts that imply non-linear system behaviour. This understanding is important for the conceptualisation of community resilience, here, with a focus on climate change in coastal areas. The community perspective, with a background in complexity theory and social emergence, bridges to the concept of social capital.

3.1.1 Emergence in Complex Systems

To provide a better understanding of the development and role of social features of interaction within communities and with their environments, the theoretical concept of emergence is insightful. In the following sections, I first examine complexity theory in depth before specifically defining social emergence and connecting it to adaptation.

3.1.1.1 Complex Systems

An important feature of social-ecological systems is their complexity. Complex systems 'are irreducible to elementary laws or processes' (Urry 2005: 3). The processes of self-organisation 'to better interact with its environment' (Manson 2001: 410) define the system as 'more than the sum of its parts', as already phrased by Aristotle (cf. Glaser et al. 2008: 77). Thus, '[c]omplexity is not the same as simply complicated' (Urry 2005: 3); simple structures can also produce complex patterns because it is the relationships that matter, not only the characteristics or the number of elements in a system (cf. Manson 2001: 406; Ratter 2012: 92). Analysing of a system according to complexity theory is about understanding the quality rather than the quantity of 'sub-component relationships' (Manson 2001: 406; cf. Ratter 2012: 91).

The roots of complexity theory lie in general systems theory. With the aim of the 'unity of science' (Oppenheim and Putnam 1958: 3) general systems theory, or *cybernetics*, describes systems most generally as 'sets of related elements' (Becker 2012: 47). The idea behind systems theory was to formulate 'universal and generalizable models that could apply to any level of analysis' (Sawyer 2005: 15). Similarly, it was the theory of *structural functionalism* in sociology (Parsons 1951), which consists of a hierarchical and deterministic conception of systems and the analysis of subsystems and components for the function and organisation of an overall system (cf. Sawyer 2005: 12). But '[d]efining the boundaries and components of a system is problematic' (Manson 2001: 411). What is part of the system and what is the system's environment? In geography, systems are defined as open and, as such, they allow for the exchange of energy, matter, and information with their environment. The definition of a system and its environment is an epistemic and intentionally theoretical construction (cf. Egner and Ratter 2008: 17). On the one hand, a system can be defined by sets of interrelated attributes. On the other hand, it can be defined in opposition to its environment. Here, everything is part of a system and, at the same time, part of another system's environment (Egner 2008b: 139; after Luhmann 1988: 292). To better understand a system's behaviour, the environment (i.e., what surrounds and influences the system) and processes within and between the system's subsystems are important (Manson 2001: 410).

While 'there is no one identifiable complexity theory' (Manson 2001: 405), the potential of 'a shift from reductionist analyses' (Urry 2005: 1) towards complexity, for the understanding of dynamics in any sort of complex system—be it ecosystems, economic systems, or social systems—has led to what Urry (2005: 1) calls 'the complexity turn'. In contrast to considering systems to be static and processes to be linear, the main features of complex systems are their non-linearity, multi-scalarity and self-organisation. In addition, complex systems consider features like agent communication and micro-macro feedbacks/downward causation, which are the basis of the concept of emergence (Sawyer 2005).

The key element of non-linearity in complexity and emergence implies that one does not know which small action actually causes a larger-scale development. An epistemological consequence of such a perspective is that '[w]e cannot make complete, absolute or final claims about complex systems' (Urry 2005: 12). Changes to a system can occur surprisingly, and radically.

A different perspective on this matter is described by the concept of path dependence, most prominently applied in economics (Arthur and Arrow 1994). The main idea of path dependence is that 'history matters', as initial (environmental) conditions shape developments and future behaviour of communities and organisations, even in the long run, when environmental conditions have changed. These 'initial actions, perhaps insignificant ones, do put us on a path that cannot be left without some cost' (Liebowitz and Margolis 1995: 2). Thus, besides producing 'increasing returns', path dependence can also imply lock-in, with undesirable consequences in a context of adaptability and environmental change (Liebowitz and Margolis 1995).

3.1.1.2 Defining Social Emergence

This leads to the question of the extent to which approaches of complexity are applicable to social systems. Social agents are different from other agents in complex systems, with consequences for the analysis of resilience and adaptability. According to Sawyer (2005: 26), social systems 'are radically more open than other complex systems'. Another difference is that social components (i.e., individuals) have 'interpretations, meanings, and intentional action' (Sawyer 2005: 26) in contrast to physical elements. How this is reflected in social emergence will be explained in the following section.

Sociologists have prominently used the concept of systems to explain society. Durkheim stated that '[s]ociety is not a mere sum of individuals' (Durkheim 1976: 422), an understanding that was upheld by later economists and sociologists, like Wundt, Hayek (cf. Sawyer 2005: 34) and Coleman (2000: 19). Accordingly, sociology seeks to understand how society is produced by the interactions between individuals. Sociologists have traditionally relied on the concept of systems when describing the notion that social interactions give rise to societies. Yet the specific definition and understanding of social emergence used in this study is heavily based upon complexity theory and has been developed by scholars from different disciplines over decades.

The literature on emergence is divided into two approaches: one following methodological individualism (explaining social phenomena by individual action and behaviour) and one following methodological collectivism (explaining collective social phenomena preceding individuals) (Sawyer 2005: 10). In social sciences, approaches following methodological individualism often use computation based on rational choice theory. However, according to Sawyer (2005: 24), social systems feature a degree of complexity that makes this method impossible. On this point, I regard emergence as a key characteristic of complex systems in a non-reductionist, collectivist understanding. Accordingly, one can even speak of 'emergent systems' that 'are complex dynamical systems that display global behavior that cannot be predicted from a full and complete description of the component units of the system' (Sawyer 2005: 102). Durkheim justifies the perspective of methodological collectivism in sociology by referring to psychology:

> Each mental condition is, as regards the neural cells, in the same condition of relative independence as social phenomena are in relation to individual people [...]. Those, then, who accuse us of leaving social life in the air because we refuse to reduce it to the individual mind have not, perhaps, recognized all the consequences of their objection. If it were justified it would apply just as well to the relations between mind and brain. (Sawyer 2005: 106).

Thus, relationships and interaction produce emergent behaviour, which is 'new and qualitatively distinct' (Egner 2008a: 40), and cannot be explained by the individual attributes alone. Manson (2001: 409) calls this 'aggregate complexity', which 'attempts to access the holism and synergy resulting from the interaction of system components'. As already mentioned, emergent behaviour results from the interaction of so-called *agents*. In contrast to an *actor*—a 'superordinate concept' (Ratter 2012: 98)—'[a]n agent is a unit or element of the system which acts locally and which cannot directly influence the system globally'. In complex social systems the agents (i.e., individuals[1]) can be aware of a global social structure (Sawyer 2005: 217) while their knowledge will always be limited (Ratter 2013: 6). At the same time, however, a resulting emergent social structure also influences the individuals' interactions. These two key features are called micro-to-macro emergence of social phenomena and downward causation (Sawyer 2005: 72) (Fig. 3.1). According to Durkheim, emergence produces 'social facts', which 'constrain individuals but at the same time emerge from the actions and interactions of those very same individuals' (Sawyer 2005: 114). Thus, instead of suggesting that individual facts can be separated from social facts, complexity and emergence assume reciprocal causation, since individual facts are actually a constituting part of society.

Different kinds of relationships (e.g., strong or weak) define the character of a system's structure. Individuals can be part of different sub-systems (e.g., family, club, community, professional network). The association with a certain subsystem, however, is not fixed, but dynamic. In social systems, the idea of emergence through relationships can be illustrated by people working together to achieve better outcomes than they would be able to achieve individually.

The analysis of social emergence critically relies on a collectivist perspective to cover the understanding of emergent social facts. In his social emergence paradigm, Sawyer divides these emergent social facts into social currents and more rigid social facts, i.e., *ephemeral* and *stable emergents*[2] (Sawyer 2005: 117). The emergence paradigm (Table 3.1) can be understood as a response to concepts of emergence that 'make too large a jump from the individual to the structural level' (Sawyer

[1]While in the context of social-ecological systems the term *agent* is common, as defined above, I will, here, use the term *individual* synonymously, as it conforms more with the further sociological language used in the context of social capital theory, too.

[2]While *emergence* describes the phenomenon of social facts emerging from lower level interaction, *emergents* are the higher level outcomes—social properties (Sawyer 2005: 2).

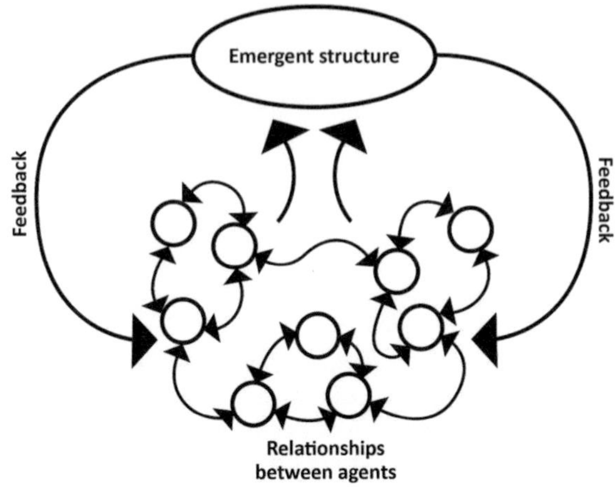

Fig. 3.1 Emergence in a complex system after Langton (in Lewin 1999: 13)

Table 3.1 The social emergence paradigm (Sawyer 2005: 211)

Social structure	Level E
Written texts, material systems and infrastructure	
Stable emergents	Level D
Group subcultures, slang and catchphrases, conversational routines, shared social practices, collective memory	
Ephemeral emergents	Level C
Topic, context, institutional frame, participation structure, relative role and status assignments	
Interaction	Level B
Discourse patterns, symbolic interaction, collaboration, negotiation	
Individual	Level A
Intention, agency, memory, personality, cognitive processes	

2005: 211). At the basis of the emergence paradigm is the individual (A), whose behaviour depends on his specific intention, agency, memory, personality and cognitive processes. The interaction between individuals on the second level (B) can be described in terms of discourse patterns, symbolic interaction, active collaboration and negotiation. The following levels take the step from the individualist towards the collectivist analytical sphere. Interaction results first in ephemeral emergents (C)—topics, contexts, institutional frames and participation patterns, as well as assigned roles and status. The reproduction of such ephemeral emergents can result in stable emergents (D), such as subcultures, linguistic patterns, shared routines and practices or collective memory. The latter two levels bridge the gap between individuals and the emergent social structure (E), which is apparent in written texts, as well as material systems and infrastructures.

My analytical focus will lie among levels B–E, with the aim of understanding how these relate to adaptation to environmental change and especially how emergents shape or confine collaborative action. Level A, agency, understood as an individual feature, is certainly important but not within the scope of this work, which concentrates on the collective side of social emergence. Social emergence, therefore, is understood as the phenomena of unpredictable and irreducible higher-level social facts emerging from lower level interactions.

3.1.1.3 Emergence and Adaptation

What are the consequences for societal adaptation to climate change, when the notion of non-linear emergence implies that 'prediction becomes impossible' (Schmid 1992)? The analysis of complex systems can help explain the effect of certain elements' or agents' behaviour on the system as well as the effect of changing environmental conditions on these agents. Within the context of adaptation, non-linear emergence turns planning into 'the management of uncertainty' (Ratter 2012: 101). Urry (2005: 3) compares self-organisation to 'walking through a maze whose walls rearrange themselves as one walks through.' These self-rearranging walls could be understood as the changing environmental conditions the system adapts to. While uncertainty means that one cannot predict and control trajectories, one can influence and stimulate certain behaviour, on the basis of a system's past pathways (Ratter 2012: 101). As mentioned with the concept of path dependence, the understanding of current conditions and future potentials through historical patterns is important. Decisions made in the past or trajectories taken—not necessarily intentionally by individuals—can make it difficult to change current conditions and lead to lock-ins.

Furthermore, a system's history is an important analytical element, due to processes of learning, and remembering: '[r]egularly occurring external relationships encourage the growth of the same set of components and sub-systems' (Manson 2001: 410). A feature that has not been well-studied yet in connection with adaptability, but is important to consider in a context of social emergence, is 'social memory' (Wilson 2015), or more specifically in the context of my research, collective experience of 'regional geohazards' (Ratter 2013). Ratter defines hazard 'as a potential threat to humans and their belongings caused by an extreme event' (Ratter 2013: 9). In contrast to risk, hazards are 'vague' and 'non-calculable' (Ratter 2013: 9). As such, 'hazard' is the appropriate term to use in the context of complex systems. Moreover, a regional geohazard is 'the regional impact of an uncontrolled risk arising from a natural feature with direct impact on the regional population' (Ratter 2013: 9). Such regional geohazards can be slow or rapid onset events.

Social memory and geohazard experience can be important for adaptation to environmental change because they can produce shared rituals and practices of environmental management or disaster response. These rituals and practices contribute to a community's resilience because a '"communal memory" acts as the sum total of learning processes embedded in communities in the form of traditions, rites and local "policies" often passed on informally and orally' (Wilson 2015: 253). In the language of social emergence, such '[h]igher-level regularities are often the result of simple rules and local interactions at the lower level' (Sawyer 2005: 3). Thus, in the context of environmental change it is necessary to identify how such rules impact society, how they are translated into collective behaviour and which processes promote or confine their emergence. An important part of that task is to consider contexts and feedbacks. Connecting social and ecological systems is important, because 'emergent properties are never purely 'social' and the kinds of processes that generate them are also not simply social' (Urry 2005: 7).

3.1.2 Resilience, Adaptability and Transformability

The concept of resilience, deeply rooted in complexity theory, provides a theoretical perspective for understanding adaptation to environmental change. Walker et al. (2004: 2) define resilience as 'the capacity of a system to absorb disturbance and reorganize while undergoing change so as to still retain essentially the same function, structure, identity, and feedbacks'. The essence of *resilience thinking*, as Walker and Salt (2006) or Folke et al. (2010) call it, is a rather qualitative question, which deals with systems dynamics. Systems around the globe can look and function completely differently. Due to the spatial and temporal dynamics, and the non-linearity of complex social-ecological systems, approaches to increasing resilience not only vary by location but also by time. The implication is that there be no universal solutions for building resilience.

3.1.2.1 Complex Social-Ecological Systems

So-called social-ecological systems (SES) are composed of a natural and a social subset, which overlap in a hybrid area. According to the definition by Glaser et al. (2012: 4), '[a] social-ecological system consists of a bio-geo-physical unit and its associated social actors and institutions. Social-ecological systems are complex and adaptive and delimited by spatial or functional boundaries surrounding particular ecosystems and their problem context'. Adger (2006: 268) asserts that '[t]he concept of a social-ecological system reflects the idea that human action and social structures are integral to nature and hence any distinction between social and natural systems is arbitrary'.

The analysis of complex social-ecological systems for the adaptive management of ecosystems (Becker 2012: 37) is most prominently institutionalised by the Resilience Alliance. Questions here are: how do societies self-organise to survive or keep functioning? How does emergent behaviour and self-organisation work as responses to changing environmental conditions? More concretely, how can this understanding help to adapt to climate change and its impacts despite uncertainty? From a complexity perspective 'living with uncertainty is part of our daily life and that surprises are only surprising because our perspective on system trajectories is basically linear and non-dynamic' (Ratter 2013: 3). Thus, resilience involves understanding possible trajectories and possibilities for adapting to changes within the current state of the system. Whether a system is in a desirable or an undesirable state, however, is clearly a normative conception.

A change in a system state can occur by surpassing certain thresholds. Thresholds 'represent the boundaries around a system state, which if crossed represent the transition into another' (Nelson et al. 2007: 401–402). According to complexity theory, it is difficult to identify and recognise thresholds, as they can become apparent or be surpassed suddenly and surprisingly (Nelson et al. 2007: 402). In that sense, thresholds can be recognised retrospectively. The idea of resilience, however, does not imply conservatism or passivity. On the contrary, a complex system state depends on the constantly changing environment as well as its non-linear behaviour characterised by feedbacks and rapid changes (Manson 2001: 410; Glaser et al. 2008: 77).

3.1.2.2 Adaptability and Transformability

An important element in understanding resilience is the distinction of the four different aspects of resilience, which are often interrelated (Walker et al. 2004). Understanding these is critical to understanding capacities to influence resilience (adaptability), as will be shown later. The four aspects of resilience are: latitude (L), referring to the number of changes a system can cope with before reaching a threshold; resistance (R), meaning how resistant a system is towards changes, or in other words, how difficult is it to alter the system's features, (also depicted as steepness); precariousness (Pr), the distance of a system to a threshold to a new basin of attraction (the closer a system is to the threshold, the lower the resilience, and the less change is necessary for transition); and panarchy (P), a system's embeddedness in dynamics of change, and adaptive cycles on lower and higher scales. How resilient a system is, therefore, also depends on factors that cannot be influenced from within the system, either in a negative or a positive aspect; processes within adaptive cycles on other scales can also lead to increased resilience of a system through cross-scale interaction (Walker et al. 2004: 3) (Fig. 3.2).

The identified aspects of resilience raise the question about the potential capacity to influence resilience. The capacity to increase resilience can imply, on the one hand, the ability to adapt to change (i.e., adaptability) or, on the other hand, the ability to transform into a new, more resilient system state (i.e., transformability). Adaptability or adaptive capacity,[3] therefore, is the 'capacity of actors in a system to influence resilience' (Walker et al. 2004: 3). In reference to the above-identified aspects of resilience, the questions arise: how can one increase the amount of

[3]As the terms adaptability and adaptive capacity are often used as synonyms, I will use the term adaptability, as in Walker et al. (2004).

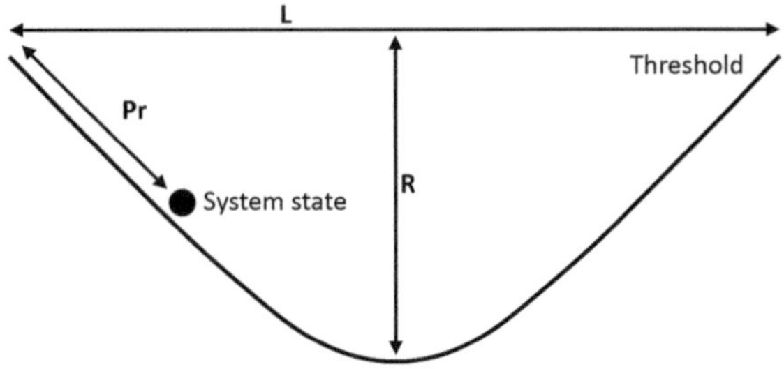

Fig. 3.2 Aspects of resilience (after Walker et al. 2004: 4)

change a system can cope with (L)? How can one increase the system's resistance (R) towards specific changes? How can one move farther from a certain threshold (Pr)? How can one influence dynamics on other scales (P)?

Furthermore, resilience can be divided into general and specified resilience. General resilience refers to the whole system and its response to any kind of stressor that could affect the system's character or way of functioning. Specified resilience refers to a system or certain part of the system, and its response towards a specific stressor (Folke et al. 2010). In the general analysis of social-ecological systems, resilience refers to the whole system—in my research, however, the focus lies on the social subsystem and the hybrid sphere with sea-level rise as a specific pressure (cf. Chap. 2).

In the context of social-ecological systems and discussions concerning global changes, the question of whether the state of the system and the changes associated with adaptation are desirable or not is a social one. In general, the global changes we are observing, such as climate change, imply great risks to our societies. However, it is a social challenge to adapt to changes associated with the pressures of climate change. Consequently, 'social change is essential for SES resilience. This is why we incorporate adaptability and the more radical concept of transformability as key ingredients of resilience thinking' (Folke et al. 2010: 2). In contrast to non-human systems, adaptability of social systems depends (in addition to internal component interaction and external environmental stressors), on features such as intention, decision-making, relations of power and cognitive elements of agency, which are shaping emergent system behaviour.

Adaptability aims at retaining the characteristics and identity of a system confronted with change. However, in situations where it is not possible to adapt, transformation becomes relevant. Transformability describes the capacity to undergo 'a fundamental alteration of the nature of a system once the current ecological, social, or economic conditions become untenable or are undesirable' (Nelson et al. 2007: 397). A society relocating from their island due to the risks of sea-level rise and transforming to a new system state in a different place with different geographical and socio-economic characteristics could be an example of transformation. In a normative sense of resilience, on a global scale adaptability is more relevant than transformability, since it would not be desirable to transform the global social-ecological system to a completely different one. On the regional scale, however, to be capable of transforming the character of a vulnerable system might be necessary. For example, on coral atolls threatened by rising sea levels, adaptation is not a long-term solution.

3.1.3 Coastal Community Resilience

Community resilience refers to the relevance of social factors for adaptability. As explained above, I apply the concept of resilience to the capacity of social-ecological systems to deal with change, focusing on social interactions and emergence and how these potentially contribute to adaptability, i.e., the capacity to influence resilience.

3.1.3.1 Community Resilience

In line with the definitions of adaptability and resilience of complex social-ecological systems presented above, Magis (2010: 402) defines community resilience as 'the existence, development, and engagement of community resources by community members to thrive in an environment characterized by change, uncertainty, unpredictability, and surprise.' By combining indicators from resilience in psychology/health and disaster research, Berkes and Ross (2013) developed a concept of community resilience that shows how place-based communities can strengthen their resilience. Place-based communities are communities in which people's livelihoods depend on the well-being of the environment, in contrast to communities in which people's most relevant ties and relationships are with a different geographic area to the one they live in (cf. Maida 2007; Berkes and Ross 2013: 13). These communities might not have any alternative to moving away in the event of an environmental change. Therefore, according to Walker and Salt (2006: 108) they deal with their environments in a more sustainable way.

As pointed out above, the resilience of social systems relies on different assumptions than natural systems because of the specific attributes of social agents, or individuals, which lead to social emergence. Berkes and Ross (2013: 14), accordingly, summarise factors that can strengthen a community's resilience by promoting its agency and capacity for self-organisation, including values and beliefs, knowledge, skills, learning, leadership, people-place relationships, a diverse and innovative economy, community infrastructure, positive outlook, engaged governance and social networks. In the context of this research, I refer to collective agency of social movements, classes, nations or enterprises. Agency, here, is understood as 'capacity for action' (Ling and Dale 2014: 4), i.e., 'the ability to [intentionally] affect events outside of one's immediate sphere of influence' (Ling and Dale 2014: 6). Agency can also be described as a 'condition of activity rather than passivity. It refers to the experience of acting, doing things, making things happen, exerting power, being a subject of events, or controlling things' (Hewson 2010: 13).

In addition to the identification of factors influencing communities' adaptability in general, the concept of community resilience is most meaningful if it is specific about who is resilient towards what, and how this affects adaptability. On the micro-level individual capacities affect the overall resilience of a community, and vice versa. However, according to the concept of panarchy, place-based communities are also embedded within larger cycles (e.g., by economic, or political ties) (Berkes and Ross 2013: 15; cf. Holdschlag and Ratter 2013). Still, increased or decreased resilience (e.g., in macroeconomic terms, whether nationally or globally, etc.), does not automatically increase or decrease resilience on the micro (community) level. Accordingly, political measures to increase resilience on the national level do not necessarily increase resilience on the local level, and can even decrease it. Moreover, the general resilience of a social-ecological system can also not be favourable to certain people, communities or institutions, faced with specific problems.

Focussing on coastal communities and particularly small islands in the context of climate change and sea-level rise, specified community resilience refers to various issues, including rapid and slow onset natural hazards. The former include storm surges, hurricanes and tsunamis, and the latter refer to events such as erosion, change in surface temperatures and shifting ecozones. Often it is a combination: e.g., sea-level rise implies a higher risk of flood events. What does this mean for community resilience then? Both rapid and slow onset events are accompanied by uncertainty in temporal and spatial terms. Rapid onset events are, by definition, surprising events, but knowledge about the changing likelihood of them occurring and understanding of past means of emergency response can nonetheless contribute to specified community resilience. While slow-onset events seem to imply better conditions for adaptation planning, they often do not develop in a linear way and are associated with changes leading to non-predictable impacts, especially due to the difficulty in defining thresholds.

Thus, community resilience in a context of climate change needs to include elements that promote mitigation of, adaptation to and recovery from potential impacts of natural hazards. The U.S. Indian Ocean Tsunami Warning System Program (2007) suggests a framework for community resilience, specifically in coastal areas, including the following elements (Fig. 3.3): effective governance, productive economy and livelihoods, effective management of coastal resources, effective land use planning, sound knowledge and management of risks, ability to receive and respond to warnings, emergency response capacity and disaster recovery systems (see also 2013a: 312). These eight elements should increase resilience towards various environmental pressures specific to coastal communities by increasing the communities' capacity to respond to both slow and rapid onset events. The framework, however, focuses especially on governance and administrative structures, while the informal level of society is relatively neglected.

Social features like community spirit and trust, social memory and collective action increase the capacity of communities to respond to environmental pressures, with social networks serving 'as the web that tie together the adaptive governance

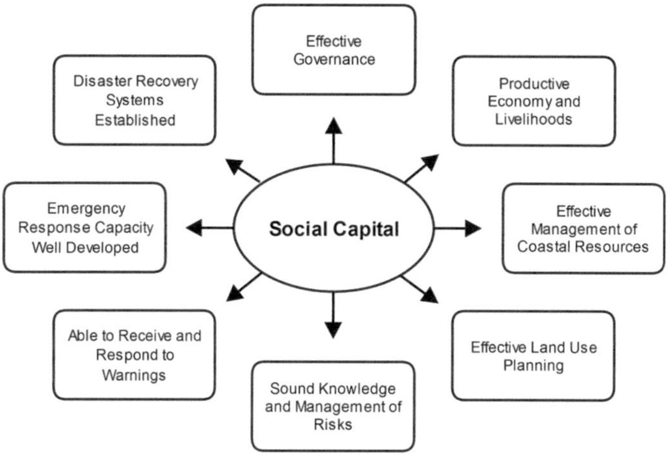

Fig. 3.3 Framework for coastal community resilience and social capital (after U.S. Indian Ocean Tsunami Warning System Program 2007: 3; cf. Hay 2013a: 312)

system' (Folke et al. 2005: 463). Adger et al. (2009), however, describe how other social factors, such as ethics, knowledge, perception of risk and values for places and cultures, can also be limiting factors for communities' capacity to cope with change and have to be considered from place to place. Pelling and High (2005) concentrate on the relevance of social networks for resilience, stating that 'implications for development policy under climate change are radical' (Pelling and High 2005: 317). Informal social relationships, trust and social norms facilitate sharing of knowledge and individual resources within the community, and vice versa. As such, adaptability for place-based communities expands not only through institutionalised social networks and co-management approaches, but also through incidental features of social capital.

3.1.3.2 Conservative Resilience

The use of resilience as a concept by social scientists has also evoked criticism. Resilience, similar to sustainability, has not only developed as a key concept for dealing with 'the environmental problem', it has also become a term that is often applied without any analytical depth. MacKinnon and Derickson (2013) summarise the main points of critique. While this critique does not specifically refer to resilience as understood from a perspective of complexity, I consider it nonetheless important to point out the risks of applying the concept in a conservative and purely normative way.

Resilience is often interpreted, conservatively, as avoiding change and remaining in a certain state that has been defined as desirable by certain actors in a community, neglecting existing power relations and inequalities. In such a context, the concept's 'regressive effects' can be explained by 'its entanglement in neoliberal modes of governance' (MacKinnon and Derickson 2013: 266–267). Both governments and environmental groups claim to define resilience and how a community must be developed for it to be resilient. Moreover, they apply the label 'resilience' to local communities, in which external influences and pressures are portrayed as fixed. In serving capitalist processes of accumulation, communities must adapt and subordinate themselves to global dynamics, a process that includes the reinvention of places. Who is resilient to what, and who carries the costs of adaptation or transformation? MacKinnon and Derickson (2013: 256) argue that the concept of resilience is used normatively especially in social sciences, where social resilience is defined, e.g., as 'the ability of communities to withstand external shocks to their social infrastructure' (Adger 2000: 361). This can lead to moral interpretations of resilience and the 'idealisation of traditional communities, which were often a place of oppression through intolerance of difference' (Robinson and Carson 2015: 5).

In fact, Weichselgartner and Kelman (2015: 263) identified that 'too many resilience-building activities draw upon unchallenged assumptions about the social world, effectively imposing a technical-reductionist framework upon more complex webs of knowledge, values and meaning—and thus action'. They highlight the importance of integrating both 'environmental and societal contributions to resilience without neglecting the differences between environmental and societal characteristics' (Weichselgartner and Kelman 2015: 263). The definition of resilience and other key terms, such as adaptability, within a context of complexity is crucial because otherwise 'there is an inherent danger that the term becomes an empty signifier that can easily be filled with any meaning to justify any specific goal' (Weichselgartner and Kelman 2015: 249).

In summary, with resilience understood in a context of social-ecological systems, strategies of adaptation will necessarily have to change due to these systems' complex, non-linear nature (Robinson and Carson 2015: 5). Linking the concept of community resilience back to adaptability, the question is how local capacities can be activated by people's agency, considering that '[t]he processes by which this occurs have not been well explored' (Berkes and Ross 2013: 15). An exclusive focus on political institutions, strategies and formal social networks neglects the relevance of less formal features of society, such as traditions, social memory and trust. Thus, it is crucial to consider the local and regional context, geography and social history when identifying collective features that shape adaptability and understanding the social processes behind them.

3.2 The Concept of Social Capital

One step towards understanding social emergence is the analysis of recurrent social practices. I seek to take social capital as an approach to addressing the 'need to couple complexity-based research with other conceptions of human agency and the role of social phenomena such as institutions and culture' (Manson 2001: 412).

Within social capital literature, one can find similar perspectives of both methodological individualism and collectivism. By identifying connections to the concept of emergence, social capital theory may provide a conceptual basis for explaining how social relationships produce society and contribute to adaptability.

The concept of social capital returns to the basic question of what holds societies together. Not only social scientists agree that 'relationships matter' and the concept of social capital has 'taken off like wildfire' (Field 2008: 1); also organisations, including the World Bank, consider social capital a key factor in poverty reduction (Krishna and Shrader 1999). Researchers apply it to explain why some societies develop more effectively than others (Narayan and Pritchett 1999; Serageldin and Grootaert 2000). In the context of resource management and environmental (climate) change research, it is considered to have explanatory power over how social networks and collective action can increase community resilience (Adger 2003; Pelling 2011).

Where does this concept come from? It is important to uncover the term's various roots to reveal the full depth of its meanings in contemporary contexts. Social capital derives from sociology, where its theorisation of civil society and social bonds can be traced back to thinkers such as de Tocqueville, Durkheim and Tönnies. Their perspective is closely connected to social emergence, though not explicitly viewed from a perspective of complexity. De Tocqueville (2002[1835/40]) pointed out the importance of civil society to democracy in his work *Democracy in America*. By distinguishing between organic and mechanical solidarity, Durkheim shifted the focus on what we would today call social networks, which make up a society that 'does not become a jumble of juxtaposed atoms [...]. Rather the members are united by ties which extend deeper and far beyond the short moments during which the exchange is made' (Durkheim 1933: 226, in Field 2008: 13). To explain how societies work and what holds them together, Tönnies (2011[1887]) distinguishes between *Gemeinschaft* and *Gesellschaft*. The former refers to bonding ties based on personal beliefs, feelings of togetherness and striving towards a common goal. The latter, on the contrary, is associated with impersonal and indirect ties, as an instrument for the achievement of individual goals (Tönnies 2011[1887]: 244).

Social capital theory distinguishes itself from these ideas, among other ways, through the link it lays between the different scales, from the individual, to the community, to institutions (cf. Field 2008: 8). In the next section, I describe the conceptual background and definition of social capital. The second section highlights modern approaches to social capital, which vary considerably, especially in their applications to specific contexts, such as development, health and environmental change. Finally, I identify criticisms and open questions regarding social capital, particularly relating to social emergence and adaptation.

3.2.1 Conceptual Background

The foundations of social capital as an explicitly defined concept lie in the 1980s and 1990s, especially in the works of Bourdieu, Putnam and Coleman, each of whom presents a different perspective and motivation for studying social relationships, i.e., individualist, collectivist, rationalist/utilitarian or Marxist.

3.2.1.1 The Classics

In contrast to Putnam, both Bourdieu and Coleman stress the utilitarian and individualistic characteristics of social capital rather than its collective character. Bourdieu takes a Marxist perspective, seeking to explain inequalities in terms of economic differences and access to means of production, rather than portraying them as results of different natural conditions. He considers social capital as the sum of potential resources made available to an individual through their membership in social networks. Individuals use these resources to maintain their status within society or even to advance to a higher status. This helps Bourdieu to analyse distinctive social class behaviours and inequalities. In this context, social capital is complementary to and transformable into other forms of capital, such as economic and cultural capital (Bourdieu 1983). Through specific social capital, people from a specific class and 'habitus' manage to achieve higher returns on their investments of cultural or economic capital than people from a lower class would. Thus, in Bourdieu's understanding, 'social capital functions to reproduce inequality' (Field 2008: 19). Is social capital actually so exclusive to elites though (cf. Field 2008: 20)?

Coleman views social capital from a perspective of methodological individualism and rational choice theory. In contrast to Bourdieu, though, Coleman sees social capital not as a resource for elites only, but rather as a resource for any individual. The idea of reciprocity, aligned with economic theory, accounts for why people cooperate even when their actions are completely utilitarian. As such, Coleman (1988, 2000) sees social capital as a combination of the individualistic and collectivist approach to society, in that it facilitates action for mutual as well as individual benefit, especially through the connection of social capital and human capital (for example, an individual benefiting from the access to resources through the family environment and community one is brought up in). Similarly, according to the livelihood approach, social capital is considered complementary to other forms of capital (human, physical, cultural capital) (Scoones 1998; Ostrom 2000). Like physical and human capital, social capital consists of 'resources that [society members] can use to achieve their interests' (Coleman 2000: 19). However, social capital is not only an individual resource but also for the collective; it appears as 'a by-product of activities engaged in for other purposes' (Coleman 1994: 312). In this sense, purely utilitarian conceptions define social capital too narrowly.

Probably the most often-cited definition of social capital comes from Putnam et al. (1993: 167), who defines it as 'features of social organization, such as trust, norms, and networks, that can improve the efficiency of society by facilitating coordinated actions'. Putnam used the concept to analyse regionally specific performances of democratic institutions in Italy (Putnam et al. 1993) and the development of civic engagement in the United States (Putnam 1995, 2000). His highly empirical work has greatly influenced policy-makers. The research in Italy analysed administrative structures and efficiency, as well as levels of trust and participation in clubs among the population. He found major regional differences in terms of administrative performance and correlations with civic engagement. This could be explained by historical analysis, which suggests a certain path dependence (cf. Field 2008: 43). Cooperative initiatives and strong connections among different businesses in medieval times, for example, are recognisable in the structure and dimensions of social relations today in the form of social capital. Putnam's later work *Bowling alone* (Putnam 1995, 2000), on the other hand, deals with the decline of social capital, decreased social life and the resulting loss of trust, as well as decreased efficiency/productivity in the United States. He identified features such as television and a generational shift towards decreased cooperation and civic engagement as causes for this.

In general, therefore, Putnam sees social capital not only as an individual and utilitarian but also as a collective resource with incidental outcomes and benefits for communities and institutions. Accordingly, Mohan and Mohan (2002: 192) consider social capital 'as a public good, to which all residents of an area have access, in contrast to social networks, which almost by definition rely on exclusion'; or, as Hunka and Groot (2011: 40) observe: '[f]rom the communitarian perspective, collective social capital is the quintessence of a society'.

3.2.1.2 Conceptual Refinement

Although the term *capital* is derived from economics, it is never used in a strictly economic sense; rather, it is used as an analogy for investment in, accumulation and use of social relationships. Arrow (2000: 4), however, critiques the use of the term capital because '[t]he essence of social networks is that they are built up for reasons other than their economic value to the participants'. Pelling (2011: 4) points out a major distinction of social capital in its 'tendency […] to grow as it is used' in contrast to other forms of capital. Most social scientists see networks and relationships as mainly beneficial to people, either to the individual, to a group of people or to the community in general. Benefits can be the product of intentional investment in social ties or of incidental outcome of such ties. To distinguish between the different conceptions of social capital and fill the term with meaning to avoid merely using it as a buzzword, one must first examine the different elements and types of social capital. Otherwise, the concept is devoid of analytical value, a problem also faced by the concept of sustainability and even resilience (cf. Sect. 3.1).

Generally, it is possible to distinguish between cognitive and structural elements of social capital (Krishna and Shrader 1999) (Table 3.2). The former describes features such as norms of reciprocity, solidarity and levels of trust within a society. Structural elements of social capital, on the other hand, consist mainly of so-called 'networks of civic engagement' (Putnam et al. 1993: 174). Arrow (2000: 4), for example, suggests that 'membership in associations strengthens political and

Table 3.2 Elements of social capital (after Krishna and Shrader 1999: 9)

Cognitive	Structural
Values	Organisational structure
Norms	Collective decision-making
Behaviour	Accountability of leaders
Attitudes	Collective action and responsibility

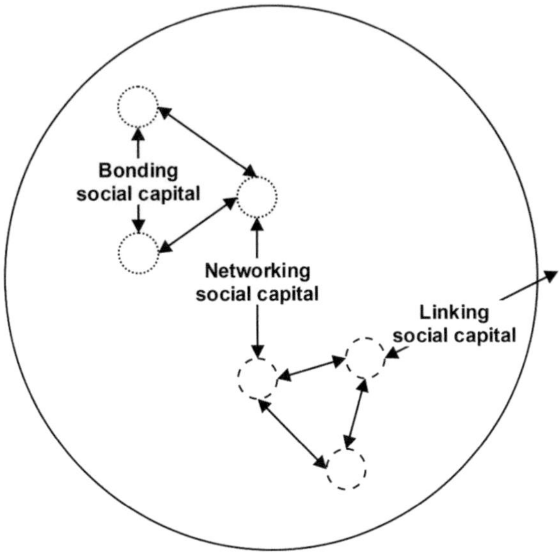

Fig. 3.4 Bonding, networking, and linking social capital [after Adger (2003: 392) and Woolcock (2001: 13)]

economic efficiency even though the associations themselves play no role in either the polity or the economy'.

A way of categorising different types of social capital is by distinguishing between bonding, networking and linking social capital (Woolcock 2001) (Fig. 3.4). Bonding 'refers to relations among family', while networking refers to 'more distant friends, associates and colleagues [...] implying connections between people who share broadly similar demographic characteristics' (Woolcock 2001: 13). Linking social capital features '[t]he capacity to leverage resources, ideas and information from formal institutions beyond the community' (Woolcock 2001: 13). It is further divided into horizontal and vertical networks (Putnam et al. 1993: 173). Horizontal networks can include bonding ties (e.g., family relationships), as well as networking ties (e.g., clubs and political organisations), whereas vertical networks usually link 'unequal agents in asymmetric relations of hierarchy and dependence' (Putnam et al. 1993: 173). Thus, social capital is also considered to have not only beneficial but also negative effects on individuals and societies, as it can also be associated with exclusion and exploitation (Ostrom 2000: 176; Häuberer 2011: 39). Structural elements of social capital are usually measured by analysing membership in various types of associations and participation in civil society organisation, as well as by accounting for individual friendship and professional networks (cf. Krishna and Shrader 1999; Grootaert and van Bastelaer 2002; van der Gaag and Snijders 2005; Claibourn and Martin 2007; Scrivens and Smith 2013).

The cognitive domain of social capital (norms, trust and values) can also be regarded as the precondition (as well as the outcome) for social networks and social capital, rather than as an actual element of social capital (Woolcock 2001: 13; Franzen and Pointner 2007; Häuberer 2011: 149; Jones and Clark 2013: 14). Accordingly, Häuberer concludes that '[t]he cultural elements generalized trust and norms of reciprocity are not social capital, because one cannot invest in them easily. Thus, we exclude them from the social capital concept and define social capital as structural entity, only' (2011: 147). However, these features are important in analysing how social capital accumulates and how it is maintained.

3.2.2 *Social Capital in Context*

The concept of social capital has gained prominence in research because of its versatility. Social capital is seen as the missing link for explanation and analysis in combination with several other theoretical approaches that involve individuals' and communities' capacity for development, migration and resource management, among others. This also makes social capital interesting for the concepts of community resilience and adaptability in a context of climate change. After providing an overview of the common applications of social capital research below, I focus on social capital theory's potential for adaptation to climate change and islands.

3.2.2.1 Social Capital Applied

The geographical aspect of social capital can be traced back to the analysis of discrepancies in regional development of institutions and societies (Narayan and Pritchett 1999; Serageldin and Grootaert 2000). It has notably been applied in human geography and development studies (Narayan and Pritchett 1999; Radcliffe 2004; Bohle 2005). The World Bank has also adopted the concept in the context of poverty reduction efforts and for policy and institutional programmes (Krishna and Shrader 1999), recognising that an overemphasis on physical capital while neglecting social capital in development policies is a cause for policy failure (Ostrom 2000: 172).

Other contexts where social capital theory has been applied are health, economic performance, migration, resource management and, importantly, resilience and adaptation to environmental changes. In economics, social capital is generally seen as a resource that facilitates access to other forms of capital, especially financial capital (Pelling 2011: 4). On the more collective level, social capital explains relative regional economic growth (Serageldin and Grootaert 2000: 44). In the context of migration, the analysis of social capital can help explain various links (for the facilitation of emigration in the first place and the integration of individuals into the wider host community). Bonding social capital becomes important for emigration as well, due to family links in host communities, but also due to the importance of remittances for the home community (Adger et al. 2002; Scheffran et al. 2012). For resource management, social capital has been suggested as an important factor enabling communities to 'overcome the free-rider problems' (Ostrom 2000: 198) associated with the so-called tragedy of the commons. In communities with higher levels of trust and cooperation, it is less likely for individuals to exploit community resources. Through norms of reciprocity, social trust can develop, which '[i]n small, close-knit communities […] [is] based on "thick trust", that is, a belief that rests on intimate familiarity' (Putnam et al. 1993: 171). Moreover, local solutions and practices of resource management based on social capital and collective action can be more efficient than top-down policies, especially in developing countries, as Kithiia (2015) shows with the example of coastal resource management in the context of climate change adaptation in East African cities.

3.2.2.2 Social Capital and Adaptation

Walker and Salt (2006: 103) stress the importance of social capital for adaptation to environmental change because, '[t]hrough people sharing and building social networks that span different areas and scale of operation, the community is, in essence, building trust and social capital that is basic to enhancing adaptability and resilience'. Social capital can remove barriers of agency and build on the individual agency for mobilisation of social capital for adaptability (cf. Ling and Dale 2014). The question is: which features of social capital specifically contribute to adaptability towards climate change on small islands? In the case of islands and climate change, for example:

> a fraction of the population only would be affected by sea level rise and would require additional protection expenditures and support, at least in most countries. This could create political and social tensions since policies would have large redistributive effects: if large investments are made, the rest of the population can see them as inappropriate; if necessary, investments are not made, the population at risk may feel unprotected by its government. (Hallegatte 2012: 3).

Therefore, 'a better understanding of local community construction could allow practitioners to utilise, support or build on local structures to enable local communities to be better prepared for flooding' (Coates 2015: 1). Features such as self-organisation and increased 'capacity for learning and adaptation' (Walker et al. 2002: 5–6), which increase adaptability are fostered by social capital by facilitating flows of information and reducing transaction costs, promoting civic engagement and public participation in decision-making processes. This is suggested to increase the exchange and processes of learning between governmental institutions and the population, and to raise public acceptance of increased costs for adaptation measures (Jones and Clark 2013). Ireland and Thomalla (2011: 5) suggest that '[g]iven the uncertainties of climate prediction and the range of potential surprises faced by communities, networks that can rapidly disseminate new and updated information are crucial in enabling effective adaptation responses'. Increased awareness of risks and opportunities is a major part of building resilience (Hay 2013a: 311). According to Adger (2003: 398) 'networking social capital and comanagement institutions enhance the capacity to adapt to the impacts of changes in climate as manifested in periodic extremes in sea-surface temperatures and gradual changes in sea level'. Berkes (2009: 1695) describes how '[b]ringing together science and local knowledge can be facilitated by bridging organizations that provide an arena for knowledge co-production, trust building, sense making, learning, vertical and horizontal collaboration, and conflict resolution'. As Ireland and Thomalla (2011: 5) observed, '[c]ollective action can also provide a space for community members to voice, discuss and solve problems'. Moreover, social capital allows for the use of local traditional knowledge, but also for the transfer of external expert knowledge and new information, which, without bridging or linking actors, might not have been accessible to the individual or community (cf. Ireland and Thomalla 2011: 5). This increases the effectiveness of collective action that is independent of governmental adaptation strategies. Thus, social capital can be seen as essential to the successful implementation of adaptation measures.

A major aspect implied by the connections between social capital and adaptability mentioned above is social capital's role as a link between society and government. As Mohan and Mohan (2002: 204) note, 'popular voluntary associations and the state have historically operated in close symbiosis'. On the one hand, this leads to approaches such as adaptive co-management (cf. Sect. 3.1.3) (Berkes et al. 2007; Berkes 2009), from an institutional perspective, while, on the other hand, the role of the individual and his or her level of information and acceptance remain in focus. Therefore, one focus of my research lies on the role of bridging organisations and analysing not only their role in the decision-making process of adaptation measures but also their impact on individual knowledge, participation and action, as suggested by social capital theory. These bridging organisations can also be described as 'nodes' that 'cut across two or more levels of organization, involve knowledge generation (or co-production), and sometimes experimentation' (Berkes 2009: 1697). Therefore, an important analytical element of social capital research is the institutional framework and the role of state and non-state organisations. Moreover, the impacts on social capital through the relationship between institutions and communities is two-way, because 'government action has an impact on people's social resources and it should be useful to take that into consideration before adopting and implementing policies' (Smart 2008: 414).

3.2.3 Missing Links in Social Capital Theory

The notion that social capital is beneficial for communities and individuals has led to a trend of focusing on the development or building of social capital. While most researchers agree on the benefits of social capital, the 'dark side' (Field 2008: 79) of social capital, which involves social exclusion, should also be mentioned. Furthermore, it is important to understand which mechanisms mobilise social capital actually to become beneficial in a certain field, such as climate change adaptation. Thus, the issue of obstacles to the mobilisation and effectiveness of existing social capital is relevant to both social scientists and policy-makers. An outline of general critiques of the concept below is followed by a conclusion of some of the open questions and research gaps of social capital theory, in as far as they are relevant to my research.

3.2.3.1 Social Capital and Exclusion

In antimodern societies, social capital can substitute for the role of institutions in case they fail to serve their purposes. However, according to Rose (2000), this can then also hinder overall modernisation and economic development and exclude people who rely on formal networks and institutions. Genkin (2006) examined the role of social capital in the emergence of fascist regimes in different countries. He suggests that the simultaneous occurrence of strong horizontal and strong vertical ties can be used by authoritarian regimes to develop. Generally, a negative aspect of social capital can be that ‘not all associations are open to all’ (Mohan and Mohan 2002: 194), and, therefore, if social capital is not accessible to everybody, it might even promote patterns of exclusion. In this sense, the negative aspects of social capital can have consequences for adaptability. Group participation is often associated with a demand for conformity and strong social control, pushing independent-minded individuals to leave the community (Portes 1998: 15). Similarly, Smith et al. (2012) suggest that strong bonding ties and exclusion inhibit the influx of creative information and problem-solving capacities from elsewhere. This, however, is not an automatism and depends on contextual factors, which either hinder or promote patterns of exclusion.

3.2.3.2 Critique of the Concept

The general critique of social capital, as already indirectly reflected upon in Sect. 3.2.1, is the concept’s lack of universality (cf. Pelling 2011). Its wide range of applications does not add anything new to basic sociological debates dealing with the essence of society (cf. Woolcock 2001: 14). More specifically, a problem of the assessment of social capital is that it is only visible when it functions. If it does not function, then it is simply not recognised and not counted (Häuberer 2011: 50). This has resulted in ‘[…] the dominant contemporary use of social capital […] to describe an outcome—collective action—not the substance itself […]’ (Pelling 2011: 5). The innovative theoretical feature of social capital, however, is its connection to a certain purpose, either for an individual, as a resource/capital to be accumulated or invested in, or for a collective, beneficial to all members, through their sharing of resources, directly and indirectly.

Similar to the critique of the concept of resilience, Mohan and Mohan (2002: 203) criticise social capital as a concept applying 'a form of revisionist neoliberalism, a political response to the alleged constraints imposed by globalization and the consequent reduction in scope for state intervention'. In the context of racism, homelessness and integration, the risk lies in governments' shifting of responsibility onto communities; social capital becomes part of the 'neoliberal discourse that legitimizes the downloading of government responsibility for social issues onto citizens' (Tanasescu and Smart 2010: 116).

Ultimately the concept lays the basis for application and further development of diverse theories, though is used less as a theory itself, and more as a metaphor. Hunka and Groot (2011: 41) suggest that '"social capital" should be treated more as an umbrella term for relations of power, influence and cultural variations, and not as a universal cure for every community-based action'. I argue that the term can only produce scientific insight if it is distinguished from normative conceptions and theory development focuses on more than just vertical relationships of power but treats social capital as a multidimensional concept, shaped by cultural, historical and geographical contexts. As such, I discuss in the following sections some of the issues that offer perspectives for the theoretical advancement of social capital.

3.2.3.3 Incidental and Purposeful Mobilisation of Social Capital

A major analytical element of social capital theory that is often neglected by social scientists is the distinction between access to social capital and the actual use, or mobilisation, of social capital. Access to social capital focuses mainly on individuals, their informal and formal relationships and networks, as originally understood by Bourdieu and Coleman. However, simply because an individual or group has access to certain resources via their social relationships and network does not mean that these resources are necessarily drawn upon or relevant to a specific purpose (cf. Lin 2001; Gaag 2005: 16). In that sense, mobilisation involves not only how groups, organisations and institutions activate available social capital, but also issues of human agency and relations of power—factors often neglected in quantitative social capital studies. Mobilisation of social capital on an individual level is difficult to identify, as it depends on many factors, such as personal skills and the necessity for help (Häuberer 2011: 142).

However, social capital is not only relevant when it is specifically mobilised for a specific purpose. Pelling (2011) makes an important distinction between incidental and purposeful effects[4] of social capital when applied to climate change adaptation. Purposeful mitigation/adaptation refers to activities such as reduction of emissions and the exchange of specific knowledge for such techniques, whereas incidental adaptation/mitigation refers to the 'underlying social relations, the social

[4]While Pelling focuses on *purposeful and incidental adaptation and mitigation*, I generalise the wording with *purposeful or incidental effects of social capital.*

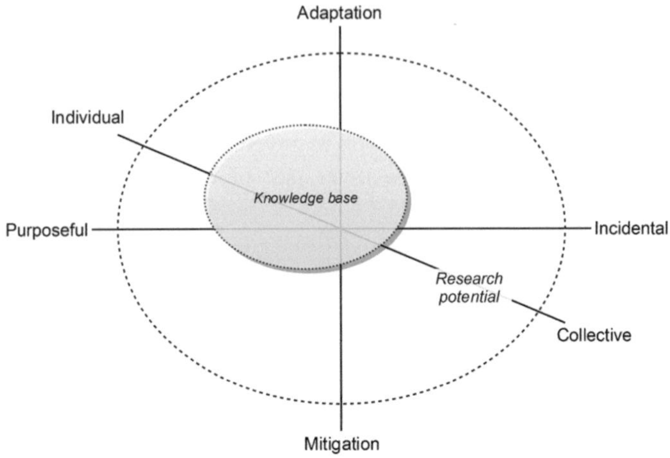

Fig. 3.5 Knowledge base and research gaps of social capital and adaptability (after Pelling 2011: 10)

raw material that determines the capacity for society to be able to mitigate or adapt to climate change in the future' (Pelling 2011: 10). According to Pelling, purposeful mitigation has been covered extensively in social research related to climate change (Fig. 3.5). In contrast, incidental effects of social capital, which are more commonly studied in contexts of health and development, have been neglected in climate change research. One can trace these effects back to Putnam's finding that communities with higher degrees of horizontal social capital, even without intentional political purpose, place greater trust in their institutions, while their institutions, in turn, function more efficiently (cf. Arrow 2000).

Most of the studies referred to in Sect. 3.2.2 make no distinction between incidental and purposeful adaptation. Often it is difficult to distinguish between the two and incidental adaptation may become purposeful while purposeful adaptation may have incidental 'side-effects' for the development of social capital (cf. downward causation). Formal and informal networks are important to both purposeful and incidental mitigation/adaptation. While analyses often assume the formal perspective, 'the informal social ties that cross-cut and often direct the formal (e.g. through corruption or social organisation at the local level that is not visible but nonetheless locally significant for reducing risk) remains difficult to capture' (Pelling 2011: 10).

Apart from the potential of social capital theory for climate change adaptation outlined by Pelling, 'the relationship between social capital as a multi-dimensional characteristic of community life and policy mitigation strategies for coastal management remains largely unexplored' (Jones and Clark 2013: 14). In fact, several authors (e.g. Krishna and Shrader 1999: 6; Naughton 2014) identified a lack of systematic analysis of how geographical and risk contexts shape 'the quality of relationships between individuals' (Mohan and Mohan 2002: 193). Acknowledging the research potential of social capital and mitigation, my research focuses on adaptation to sea-level rise, as this is the most pressing issue for small islands. My focus, however, does not lie on showing how to assess or quantify the level of social capital in an island community or among its individuals. I analyse how the various supposed effects of *collective social capital* are mobilised and contribute to community resilience to climate change. This includes the consideration of specific geographical context and the question of whether insularity results in a distinctive manifestation of social capital.

Can collective social capital be relevant to adaptability towards climate change, especially in place-based communities such as small islands? How can incidental adaptation supplement purposeful adaptation? Community resilience can only be full assessed when social capital's incidental effects on community climate change adaptation are also taken into consideration. Based on the variety of definitions and proposed features of social capital, I examine functions of social capital that are relevant to collective action and adaptation to climate change on small islands. I reject the neo-capitalist perspective of social capital as 'investment in social relations with expected returns' (Lin 1999: 30); rather I conceive of social capital as a collective good, a resource that is potentially beneficial for a whole (island) community (cf. Mohan and Mohan 2002).

3.3 Nissology

With a third theoretical approach, I integrate the study of islands into the analysis of social-ecological resilience and social capital. From a theoretical point of view, islands are interesting places for scientists due to their seemingly clear borders. But are islands—especially 'small' islands—really a 'closed system, representing a microcosm of the planet Earth' or even 'model living laboratories for the Earth' (Nagarajan 2006: 296)? When studying place-based communities in a context of environmental change, what can small islands specifically teach us about how 'place matters'? Assuming a nissological approach introduces the analysis of place-specific factors, as well as island-specific features that are relevant to the development of social capital, into the context of social ecological systems and adaptability.

Small islands have been the focus of media and public attention due to their perceived susceptibility to the pressures of climate change. However, researchers do not only study small islands because of their supposed vulnerabilities. In general, people have always been fascinated by islands, whether as an inspiration for writers, a frontier for explorers, a place of interest for scientists or a place of dreams and hopes, especially for non-islanders (McCall 1994: 1; Schofield and George 1997: 5; Rainbird 2007: 2; Mountz 2015). In fact, '[i]slanders have shown little tendency to romanticise [island living], leaving that to mainland landlubbers' (Gillis 2014: 160). Baldacchino and Clark (2013: 131) conclude that '[i]slands maintain a unique grip on our imagination as intriguing places where fantastic utopic and dystopic worlds are possible'.

The following section introduces features that make island societies, in particular, vulnerable and possibly resilient. I highlight the main concerns of social scientists regarding 'small islands. Furthermore, I will discuss main theoretical approaches to the study of islands by critically reviewing definitions and concepts of (small) islands in science and literature, and by introducing the concept of nissology—a critical way of studying islands, which is important for my further research design, methodology and interpretation of results.

3.3.1 *Defining and Understanding (Small) Islands*

The definitions of islands and the origin of the actual concept of 'island' as opposed to 'land' have consequences for epistemological and methodological approaches. Therefore, this chapter will give an overview of the various definitions and discussions about the geographical conception of islands.

The definition of islands is more difficult than it seems at first glance. The origin of the word 'island' is Old English 'igland'—'ig' from the Latin 'aqua', developing to isle by the influence of French ('île') in the fifteenth century (Royle 2007: 33; Hay 2013b: 211). On this basis, a simple, perfectly geographical definition is: '[a] piece of land completely surrounded by water' (Royle 2007: 34). However, this definition does not account for all of an island's dimensions. Where are the lower and upper limits? Can a rock or a reef be considered an island, and can an island also be a continent? Numerous terms refer to land masses surrounded by water, such as the archipelago, atoll, isle, islet, key, reef, or rock. There are river islands, islands in lakes, tidal islands and classic oceanic islands, all of which 'have such a variety of shapes, physiographies, coastlines and other geographical attributes that no meaningful conclusions can be drawn from their systematic study' (Nunn 1994: 6). Thus, perhaps 'it is not necessary to have rigid definitions', because '[…] we know instinctively what is an island […] when we see one, and that is the important thing' (Nunn 1994: 1). Yet description and categorisation from different perspectives can help to understand the diversity of islands, to study particular features that many or all of them might share and to point out similarities between islands and continental areas. These are important factors for scientific case study research dealing with islands. They help to avoid misconceptions that lead to false assumptions, interpretations or generalisations.

The broadest categorisation is the one of continental and oceanic islands. This division, however, neglects the diversity of island types within these categories as it is based solely on location. Therefore, a genetic classification of islands is insightful and important for the understanding of physiographical characteristics. Here we can distinguish plate-boundary islands and intraplate islands, which, in turn, can be further sub-categorised (Nunn 1994: 10). Another system of classification is a relatively functional one, taking into consideration the administrative status, population and economic structure of an island. Moreover, the functional classification regards isolation not in terms of geographic distance/remoteness, but rather in terms of connectedness and the impacts of isolation (cf. Klug 1985). The genetic and functional classification of islands can, therefore, combine various perspectives, which lay the basis for a more analytical comparison among the great diversity of islands worldwide.

Most definitions rank island continents as the upper limit for defining islands, while their lower limits vary widely. Furthermore, geographical insularity can change through mainland connections that can be established or destroyed by natural or anthropogenic factors (King 1993: 15) such as tide, storms or bridges and embankments (Royle 2007: 44). It becomes even more complicated when socio-economic factors are considered because 'the physical nature of islands is going to have its social and cultural influence' (McCall 1994: 3). Accordingly, it is unclear what constitutes an island community. Does an island have to be politically autonomous? Do seasonal workers, for example, belong to the island community? For many islands, emigrants, who still hold strong ties to an island (e.g., through remittances), can be an important economic and probably even political factor for the island. Due to the ambiguity of definitions and the role of people's identity as islanders, Kelman (2003: 71) even concludes that it can be up to the community members to define whether they consider themselves islanders (e.g. because of the experience of isolation and smallness, which can have various causes).

Is it possible to distinguish small and large islands, and what is the significance of such categories? From a geographical point of view, it is widely agreed that there are several aspects that make islands, and particularly small islands, peculiar (Dommen 1980: 931).

Hay recognises that:

> larger islands differ significantly from smaller islands. Islands dominated by a mountainous hinterland may have little in common with low-lying sand or coral islands. Tropical islands differ dramatically from temperate or cold-climate islands. And I do not only wish to establish these as intractable differences on the ground—the construction of island representations faithfully reflects this stubborn proliferation. (Hay 2013b: 210)

Moreover, the categories small and large can be used in relative and absolute terms: McElroy (2002: 49) considers population (less than one million) as the main factor. Hess (1990: 3) considers size as a factor in addition to population ($\leq$10,000 km^2 and $\leq$500,000 residents) for 'independent island states, archipelagic states and islands associated with larger countries'. However, an island within an archipelago can be relatively small, as can a dependent island compared to its mainland or a mainland island compared to a continent (Stratford et al. 2011). Somehow, however, only small islands seem to have a certain 'appeal' regarding 'seclusion and retreat' (Schofield and George 1997: 5), without which they would sooner resemble continents.

From a biogeographical perspective, isolation, especially of small islands, became the centre of attention of evolutionary ecology, when analysing islands such as the Galapagos (Schofield and George 1997: 7). Teaiwa (2007: 514) suggests that 'to island' is to view the earth as an island and to teach others how to live on an island, that they might learn how to live in an environment with finite resources. Nagarajan (2006: 294–295) even describes islands as 'microcosms of the planet earth' and states that 'the study of small islands, from interdisciplinary perspectives, can provide social-ecological models for larger societies, and the planet Earth'. This perception can be misleading, though, because islands often face the same problems as continental regions and sometimes even more connections to other places. The idea of islands as closed systems and the alleged border between land and sea are a construct and often an illusion (cf. Gillis 2014: 155), as will be discussed further below.

Small islands have been a particular focus in science, as well as politics, especially since the formation of networks such as SIDS and the Alliance of Small Island States (AOSIS). Both are political alliances seeking to strengthen islands' voices within the international community. SIDS define themselves as small islands 'and low-lying coastal countries' that:

> share similar sustainable development challenges, including small population, lack of resources, remoteness, susceptibility to natural disasters, excessive dependence on international trade and vulnerability to global developments. In addition, they suffer from lack of economies of scale, high transportation and communication costs, and costly public administration and infrastructure. (UN-OHRLLS 2016).

This definition, however, is not very precise as it still includes islands with a wide range of sizes, populations and GDPs (Crowards 2002: 144).

As mentioned above, the appeal of (small) islands to natural and social scientists has been attributed to geographical and ecological features and has also been explained in political and even emotional terms. Royle (2007: 52) concludes that it is impossible to 'satisfactorily' answer the question of what an island is. Nonetheless, it appears that 'island(er)s have a sufficient commonality to warrant looking at them comparatively, justifying a systematic 'island studies' perspective' (Baldacchino 2005a: 247). Seeking to avoid normative and deterministic terms, McCall (1994: 3) summarises the discussion on island features, also considering cognitive and social features (see Table 3.3).

3.3.2 A Vulnerable World of Islands

Many of the features of small islands described in the previous chapter have led to a common perception that small islands are highly vulnerable. Which, then, are island-specific features of coping with environmental pressures, i.e., 'characteristics [...] that influence their capacity to anticipate, cope with, resist and recover from the impact of a natural hazard' (Blaikie 1994: 11)? In addition to impacts from natural hazards, vulnerability here also refers to socio-economic impacts. Generally, small islands' vulnerability can be distinguished by so-called issues of scale and issues of location (cf. Kerr 2005).

Table 3.3 Shared island features, after McCall (1994: 3–5)

Clearer land boundary than continental regions and 'a sense of the sea as part of their lives, not an isolating barrier'
Sea resources, EEZ (larger than actual land mass); EEZ islands have a large area under their control by a small population
Islands often claimed by continental states for **strategic** reasons (especially when singular and remote/isolated)
Limited terrestrial resources
Boundedness to the place and **differentiation** of islanders and non-islanders (though on large islands, sub-division can play a dominant role)
Sense of smallness due to comparison (a rather (post-)modern problem)
Social relations are closer; people know each other and trust in certain ways of dealing with each other. Bureaucratic structures might, therefore, fail on small islands
Periodic and imperative **migration**

3.3.2.1 Islands' Issues of Location and Issues of Scale

Issues of location on small islands are, for example, expensive and unreliable transportation, vulnerability to natural disasters and distances to centres. Royle (2001: 43) calls the dependence on sea and air connections for the transport of goods and people the 'most obvious and basic constraint of insularity' (cf. Brookfield 1990: 25). The mostly peripheral setting of islands also complicates the transfer of innovations from core regions (Royle 2001: 45). On the biogeographical level, islands' isolation engenders a genetic uniqueness of species which is vulnerable to disturbance (cf. Moro et al. 2005). King (1993: 33) extends this logic of isolation to the social level (e.g., disturbance by mass tourism). However, positive issues of location can also be identified, such as islands' function as a refuge, as stepping stones (more historically relevant) or as strategic locations (e.g., for defence) (cf. Royle 2007: 49–54). Dommen (1980: 931) even finds that island could be considered as 'particularly fortunate places, where life is longer and nature is bounteous'.

Issues of scale include, amongst others, that the lack of a market forces a high degree of openness onto islands (McElroy 2002: 49). Limited economies of scale and a narrow range of exports lead to high specialisation and dependence on a single export product. Thus, changes in global market patterns and prices are dangerous (cf. Hess 1990: 5; Royle 2001: 61; Briguglio 2002: 73). The lack of natural resources, consequently, leads to a dependence on strategic imports such as food and fuel, as self-reliance is hardly possible (Hess 1990: 4; Walker and Bellingham 2011: 306). Issues of scale further include limited human resources and the 'limited capacity to spatially segregate activities produces tight feedback loops between any environmental stressor and its impact' (Kerr 2005: 508). Thus, according to Depraetere (2008b: 21) the limitations in land area and natural resources can lead to more interactions within the island system than externally.

Island communities encounter conflicts, or rather 'confrontations' (Depraetere 2008b: 20) of different activities more directly, such as construction and land development opposing natural conservation. The loss of already limited natural capital through either overexploitation or degradation due to external influence, or the overshoot of an island's carrying capacity by overpopulation, can lead to the collapse of the society. Examples of such developments, caused by both internal and external pressures, are described on islands such as St. Matthews Island (Walker and Bellingham 2011: 5), St. Kilda (Maclean 2006) and, as prominently discussed by Diamond (2005), Rapa Nui/Easter Island. Despite the possibility of sustaining a society that is larger than the available natural capital can support (e.g., by imports), emigration is a major issue for islands. Work opportunities and higher education are limited, especially on small islands. Therefore, the reliance on remittances supports a development where 'emigration may become institutionalized as part of island society, necessary for its stable survival' (King 1993: 23).

Regarding economic development, a common assumption is that small islands' economic problems can be compared to the general discourse on developing countries (as economic peripheries in relation to industrialised centres). Islands face similar problems to developing countries, though 'more intractable' (Byrne and Inniss 2002: 5). In fact, 'three out of four of the world's small developing countries are island states' (Douglas 2006: 75). Jesinghaus (2002: 315), on the other hand, warns to maintain the differentiation between industrialised and developing countries when discussing small islands, as there are still many differences between them in terms of wealth, autonomy and economic development and structure, even though they all have limited space and relatively long coastlines in common. A common economic characteristic of many islands is the development of an economy that shifts away from colonial exports and towards tourism and financial services: '[o]f the 31 countries in the world with +20% of their GDP generated by travel and tourism, 27 are islands states' (McElroy 2002: 49; Kerr 2005: 509). Furthermore, islands are often more densely populated than other regions with a similar degree of urbanisation (Dommen 1980: 932). In the context of small islands, this means increased stress on local resources and ecosystems. The latter are generally considered more fragile and closely linked on islands, where small changes can have a critical impact on major ecosystem functions (Hess 1990: 4; Millennium Ecosystem Assessment 2005: 27).

The Barbados Programme of Action (BPoA)[5] defined sustainability issues on islands for the first time and outlined commitments and objectives to sustainable development and monitoring by SIDS (United Nations 1994: 3–5). Amongst others, it includes the importance of self-reliance, protection and development of human and cultural resources, and the promotion of partnerships between non-governmental organisations and governments. The Mauritius Declaration of 2005 reaffirmed the BPoA as the 'blueprint for small island developing states and

[5]Also: *Report of the Global Conference on the Sustainable Development of Small Island Development States* in 1994, Bridgetown, Barbados.

the international community to address national and regional sustainable development in Small Island Developing States' (United Nations 2005: 2). In general, the promotion of sustainable development on islands should address resource management, the provision of human services, and small-scale and non-resource-dependent development options (Hess 1990). Efforts to increase sustainable development and tackle each of the issues described above focus on both internal and external factors. Endogenous approaches usually include increased self-reliance 'in terms of social and personal aspirations' (Hess 1990: 9) to decrease dependence. Sustainable tourism, amongst others, plays an important role in many islands' sustainable development strategies as it integrates an economic perspective into the sustainability concept.

Despite the initiatives and commitments outlined, the implementation of sustainable development measures on small islands has progressed only slowly (UN Secretary General 2010; Hay 2013a). In many cases, human activities have increased vulnerability on small islands towards hazards associated with global (climate) change (Campbell 2009; Hay 2013a). Most recently, the Small Island Developing States Accelerated Modalities of Action (SAMOA) Pathway reaffirmed the dedication of the international community to strengthening sustainable development on SIDS (United Nations 2014b). Similarly, SIDS drew attention at the Conference of Parties (COP) meeting in Paris in December 2015 by calling for more ambitious reduction targets for greenhouse gas emissions (cf. Bawden 2015).

3.3.2.2 Small Islands and Climate Change

The currently 'most pressing' (United Nations 2014a: 40) problem and probably the 'greatest potential threat' (Byrne and Inniss 2002: 10) to small islands is global climate change with its various consequences (cf. Chap. 2). An observed sea-level rise of 20 cm during the last century has already caused coastal land loss. In fact, the coastal vulnerability index developed by UNEP shows that all of the SIDS face either a moderate or high risk (UNEP 2005: 43). Further sea-level rise of several metres would cause major problems not only for flat SIDS, such as the Maldives or Tuvalu, with maximum elevations of less than two metres. Also dependent islands that are neglected in such statistics and continental countries with vulnerable coasts are affected by the impacts of rising sea levels. In general, distinct local geomorphological features must be considered. Atoll shores, for example, are mostly rock, not sand, and, therefore, less flexible to sea-level rise (Pilkey and Young 2009: 57).

Further consequences associated with sea-level rise include increased salt water intrusion, flooding of freshwater wetlands and valuable agricultural zones (Walker and Bellingham 2011: 298), as well as higher peak tides and increased impacts of (potentially more frequent) storm surges (Byrne and Inniss 2002: 9). Changes in precipitation patterns making dry islands drier and influencing the composition of vegetation (Walker and Bellingham 2011) exacerbate the situation that small islands already suffer from having very limited catchment areas for precipitation, causing widespread water scarcity (United Nations 2014a: 41). The interconnectedness of global changes produces risks, particularly for SIDS, on the economic, environmental, geopolitical, societal and technological levels—all affected by climate change directly or indirectly—and causes SIDS to be more vulnerable than other Least Developed Countries (LDC) (Hay 2013a).

When it comes to climate change mitigation, external factors must be addressed first. Reduction of emissions on islands alone would not avert the consequences of climate change, as islands account for merely a small portion of worldwide emissions (cf. Byrne and Inniss 2002: 15). Thus, the AOSIS emphasised its strong support for the Kyoto-Protocol and UNFCCC goals to reduce emissions. Paradoxically, the greatest potential threat—climate change—has affected the policy discourse in such a unidimensional way that other problems of sustainable development, as described above, are often neglected: 'climate change has become a distraction from underlying vulnerability and lack of development' (Kelman 2014: 126). Moreover, adaptation strategies can be misused and designed for the benefit of an influential elite, including politicians, businesses and researchers (Baldacchino and Kelman 2014). Thus, Hay (2013a: 324) suggests 'improving risk knowledge, governance, coastal resource and land use management, disaster prevention, emergency response and crisis recovery, while also strengthening socio-economic systems and livelihoods' to align reduction of vulnerability with sustainable development goals.

3.3.2.3 Island Resilience

McCall (1994: 1) criticises the widespread view of islands' vulnerability that envisions islands as 'places [...] in need of saving; as places that must be improved and brought to dominant continental standards'. Analysing disaster reduction measures, Campbell criticises the western-centric point of view on islands and vulnerability, since before colonisation 'vulnerability was not a particularly marked characteristic of island societies and communities and, accordingly, islands are not inherently vulnerable places. On the contrary, they were resilient' (Campbell 2009: 94). As Kelman and Khan (2013: 1132) observe, the 'construction of islandness as immediately equating to vulnerability is a myth. [...] [T]he same island characteristics simultaneously contribute to vulnerability and resilience'. If one examines the distinct features of small islands from a different perspective, focusing on their potentials and capacities to adapt to change rather than their weaknesses, the concept of resilience becomes a suitable analytical frame (cf. Sect. 3.1.1). What shapes islands' resilience, i.e., their ability to deal with climate change?

Humans have lived on islands for centuries. Even if they are distinct and vulnerable, islands also offer great opportunities. Islands have long succeeded in managing the finiteness of their resources, even before major contact or trade with other world regions (Kerr 2005: 504). In contrast to the arguments of the general disadvantages of their isolation, island communities used to 'engage with outsiders socially, and continuously incorporate elements of this contact into their own populace, and through contact and exchanges create a distinct community identity located historically on sea and land' (Rainbird 2007: 173). Against the background of so many constraints and of living in a constantly changing environment, islanders developers efficient traditional adaptation strategies. In fact, Turner et al. (1996: 170) suggest that '[s]ome traditional societies have developed networks for support and reciprocity that are more effective than the natural disaster programs of even the wealthiest countries'. Without romanticising traditional skills and methods, it must be argued that such resources are combinable with modern and technological strategies to tackle problems. This can contribute to the development of technologies and concepts, not only for islands but also for the entire world (Cambers 2007; Kelman and Khan 2013).

Several scholars point out the strong feeling of a common identity that can often be found among communities on small islands. Especially small islands develop high bonding social capital, which is important for the continuity of an island community otherwise threatened by emigration (Hay 2013b: 25). Such dense social networks and cooperation are an important factor for resilience (Barnett and Adger 2003; Baldacchino 2005b). Networking and linking social capital, however, are also important for islands. Campbell (2009: 91), for example, observed an increased inter- and intra-island cooperation to increase food security in traditional Pacific islands communities. Inter-island networks and mechanisms of help increase resilience (e.g., after cyclones) (cf. Kelman and Khan 2013: 1133) or through initiatives linking small islands from around the world, such as the Many Strong Voices project.[6] Life on small islands depends on this cultural and social capital. As such, Baldacchino (2005b: 32) suggests that 'small island territories may be the best sites for seeing the effects, or absence, of a strong social fabric'.

Given the complex interrelations and causes of issues on small islands and the urgency for action in the face of climate change risks, Nunn et al. (2014: 234) call for 'assistance to be targeted directly at the community level where the greatest disconnect lies between the science and stakeholder awareness of climate change'. A resilience perspective encompasses the limitations and opportunities on small islands. It helps to avoid copying or dictating strategies from continents to islands, as well as mixing up and over-prioritising particular aspects of probable vulnerability. This risk can also come from the influence of external stakeholders and institutions, and can be avoided by integrating a perspective outlined by the concept of *nissology*, as will be discussed in Sect. 3.3.3.

3.3.3 Thinking with the Island

Nissology is the approach of comparative and systematic island studies, recognising the shared features of islands as well as their diversity. It is a (geographic)

[6]http://www.manystrongvoices.org/ (checked on 12/01/2016).

perspective of islands on a global scale and a call for critical research and to study 'islands on their own terms' (McCall 1994: 1).

There are several approaches related to the ideal of nissology, suggesting alternative perspectives to the concepts of simplified constructs of islands as small and closed systems, particularly vulnerable and peripheral. Such approaches might add to the understanding of island-specific features of resilience or at least show how to design research accordingly and to potentially identify and interpret such features.

From a global perspective, islands can be seen not as peripheral or minor land masses on the globe, but the opposite: the main element in a 'world archipelago' (Depraetere 2008a: 3). Archipelagic formations—groups of islands—are very common; in fact, the world can be regarded as an archipelago. Many SIDS are archipelagos, as are many large countries (e.g., Japan, the Philippines, the United Kingdom). Others are shaped by their coastal landscape, consisting of islands (e.g., Canada). Thus, geographical features of islands are actually the rule rather than the exception (Depraetere 2008a; Stratford et al. 2011). LaFlamme (1983) identifies four attributes of an archipelago: it consists of a large number of islands; the surrounding waters are within its boundaries and integral to its heritage; most of the constituent islands are small and economically underdeveloped; and it has a centrifugal tendency including a rivalry between the centre and the other. The questions of how many islands make an archipelago and how 'scattered islands' fit in, remain open.

From a systems perspective, the goal of 'archipelagic thinking' is to overcome and look beyond two main research perspectives in social sciences: the uniqueness of islands due to the border 'land/sea' and the opposition of islands and continents (Stratford et al. 2011: 114). Islands are seen as the same, although 'no two islands are ever alike' (Stratford et al. 2011: 117). Islands are separated from archipelagos or an archipelago is unified into one island. Islands are viewed as peripheral to the mainland and are studied from a development perspective. Instead, archipelagic research seeks to compare different islands and understand them in terms of archipelagic relations. Rather than highlighting isolation, the focus here is on islands' characteristic interconnections, i.e., how islands also work together—'island movements' (Pugh 2013). Often one island cannot be understood without the other. Therefore, such research can speak of 'assemblages' and 'constellations' (e.g., the Pacific/Oceania, the Caribbean). The focus on archipelagos, however, still seems to favour land over water. In a context of research regarding people's response to sea-level rise, understanding perceptions and behaviour in relation to the sea is necessary. Rainbird (2007: 163) concludes that 'islands should be a focus of study in regard to their relations with the sea rather than the apparently circumscribed piece of land they represent' and that '"island studies" […] must have to do with water, the element common to all islands, and, more specifically, the sea' (Hay 2013b: 211). A perspective of complex social-ecological systems considers marine and terrestrial pressures affecting the natural and social subsystems of coastal and low-lying areas, as well as respective feedbacks (Fig. 3.6).

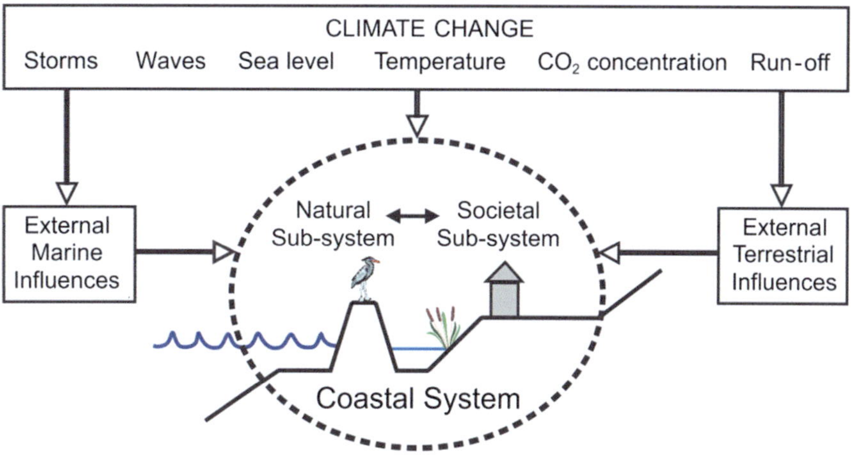

Fig. 3.6 Climate change pressures on the coastal social-ecological system (*Source* Nicholls et al. 2007: 318)

Similarly, the concept of *ecotones* represents an approach that accounts for the special relationship of land and sea on islands that distinguishes them from continents (Gillis 2012, 2014). It is derived from environmental sciences and refers to 'places where ecosystems intersect, overlap and exist in creative tension with one another' (Gillis 2014: 155). It is a suitable concept for islands, because they 'are never the clearly bounded entities we imagine them to be' (Gillis 2014: 155); the margin between land and water is a construction. According to Gills, therefore, the idea of ecotones not only refers to the ecosystem but also to society—islanders as coastal people (cf. Rainbird 2007). Although the distinction between islands and continents is a fairly modern construct, on continents, the ocean is considered more of a border, as 'land's end', whereas for islands, the sea represents the beginning of a journey. Living with the shore means understanding its function as a margin (habitable, transitional), but not as a barrier (construction, hardening, protection, straightening) (Gillis 2014: 160). Thus, issues involving coastal management, such as adaptation to climate change, must consider the different perceptions communities have towards coastal zones, which role they play for community structures and relationships and which implications they have for the implementation of adaptation measures. The concept of ecotones helps to understand the dynamic nature of coasts and suggests a way of life "*with* rather than *on* our shores" (Gillis 2012: 196, italic in original).

Apart from the question of what the object of island studies is, it is important to point out how and for what purpose nissological research is done. Pointing this out has methodological implications and it is an element of critical science. Too often island-related research in the social sciences is *about* or *on* islands and islanders, but not *with/for* them; it stresses differences, the 'other', and boundaries (cf. Stratford et al. 2011). Therefore, McCall (1994) suggests four dimensions of general scientific

Table 3.4 Principles of nissology (McCall 1994: 6)

Height	Communication of research outcomes to decision-makers, national and international
Width	Communication of results to society, including popular publications
Depth	Substantive research also of basic questions; open-mindedness towards results
Time	Provide research and results for future generations

principles for nissology (Table 3.4): height, width, depth and time. Height refers to the communication of research results to political stakeholders on various scales, to provide knowledge that can be integrated into policymaking and planning. Width refers to the communication of research outcomes to society. Thus, results should not only be published in scientific journals but should also be made accessible to the public, especially the population of the island where the research has been undertaken. Depth means to study not only specific research questions, but also to be open to dealing with general questions, especially in cooperation with fellow international scientists. The dimension of time refers to research being directed to the future so that results are relevant and accessible for future generations.

In conclusion, islands and island communities should not be viewed from a continentalist perspective, neither should focus be placed solely only on islands' vulnerability (cf. Betzold 2015). '[W]e need not feel sorry for islands and [i]sland states' (McCall 1994: 1), nor should we romanticise their resilience. To study islands on their own terms, therefore, must not imply localism or a disregard for islands in the global context. Islands have a strong maritime identity but have always and will always be also shaped by contacts and exchange with non-islanders, the same way as the island mentality also shapes continental people's lives.

Nissology and related concepts introduced above offer the perspective to conduct research 'with' and 'for' islands/islanders rather than strictly 'about' them. This approach is needed in times of climate change and social sciences are able to fulfil this need. Generally, the relatively young field of island studies and the question of what nissological research is about needs further development, inspiration and case studies. As outlined above, much of the research on islands, from various disciplines, focuses on islands' alleged vulnerabilities. Thus, a perspective of island resilience can contribute to the further development of the field.

3.4 Questions to the Island Case Study

The preceding chapters described and reviewed three different streams of theory, which are brought together here. The aim of using these different theoretical approaches is to provide an adequate framework for addressing the specific research question for this work:

How does social capital help explain collective behaviour as well as opportunities and limitations of societal adaptability towards non-linear climate change on small islands? In other words, what shapes collective adaptability to climate change in island societies?

For research of complex social-ecological systems, O'Sullivan et al. (2006: 612) highlight the importance of 'close study of the local situational characteristics of physical locations, of interactions among neighbouring locations, and of the flows along interaction networks' as 'key to understanding and anticipating its behaviour'. The theoretical perspective applied here can be summarised as viewing social-ecological systems through a social capital lens, with a focus on a specific geographical context—in this case, small islands. Further *questions to the islands* should lay the basis for the operationalisation of a research framework, which I develop in Chap. 4.

How can adaptation be understood and analysed in a context of social-ecological systems, sea-level rise and uncertainty? I analyse features of the social subsystem and the hybrid sphere, as described in Sect. 3.1.1, with environmental feedbacks and forcing from the natural system. More specifically, this analysis includes the aspect of adaptability—in this context, the role of social relationships and structures, both formal and informal, for the specific resilience to sea-level rise. Contributing to the understanding of community resilience (Sect. 3.1.3) in the context of environmental change by the theoretical framing within complexity theory, offers a focus on social emergence by methodological collectivism. Concrete sub-questions here are: how does higher-level order (island societies) shape adaptability (in this case, towards climate change) and which lower-level processes (social capital) lead to its emergence? In other words, how does social capital theory (Sect. 3.2.1) explain the emergence of a specific higher-level order (Sect. 3.1.1)?

Having identified a lack of research on the mobilisation of social capital (Sect. 3.2.3), I examine social capital on a community level as a collective good, rather than on an individual level. In avoiding reductionism, such analysis must be multi-dimensional and focus on relationships rather than on individual components. What are the social facts (collective goods) that can only result from social interaction, i.e., through social capital? The analytical focus lies on interaction and the resulting emergents, their benefits, as well as specific factors that influence emergence. Through a social capital lens, the question then becomes: how does the mobilisation of social capital increase island communities' adaptability towards sea-level rise?

On the individual level, people may act egoistically/utilitarianistically, but, according to complexity theory, 'lower elements are unaware of their role in emergent phenomena' (Manson 2001: 411) so individual actions may benefit a whole community. Thus, I particularly distinguish between incidental and purposeful social capital, producing ephemeral and stable emergents, and analyse the purpose of the most relevant elements of social capital for adaptation to sea-level rise on small islands. Formal and informal collective action can both be viewed as features of self-organisation in complex social systems (cf. Ratter 2012: 91). What are limiting and promoting factors for the effectiveness of social capital with regard to adaptation?

Finally but importantly, the nissological lens contributes to the theoretical frame by integrating the analysis of place-specific features for the development and mobilisation of social capital (Sect. 3.3.1). This does not imply following the island-continent dichotomy, but rather focusing on inter- and intra-island social features, according to the nissological approach of studying islands on their own terms (Sect. 3.3.3). If emergence is actually higher where 'components interact in densely connected networks' (Sawyer 2005: 4), the following question arises: how does a specific geographical context influence social relationships (cf. feedback, downward causation) and the emergence of social facts that influence a community's resilience in a social-ecological system? This includes questioning whether, in fact, '[m]any of the characteristic features of islands and islanders are the product of islands' ecotonal nature', as suggested by Gillis (2014: 159). As the discourses around community resilience, communities of place, and islandness in general are about people, perception, attitudes and behaviour, such an analysis can facilitate a critical understanding of the value of social capital theory for adaptation in vulnerable regions such as small islands.

The following chapter includes the operationalisation of the various questions above and describes the choice and development of adequate research tools. By interpreting the results of an empirical case study (Sect. 5.4), the questions to the islands will be readdressed in the final chapter.

References

Adger, W. N. (2000). Social and ecological resilience: Are they related? *Progress in Human Geography, 24*(3), 347–364.

Adger, W. N. (2003). Social capital, collective action, and adaptation to climate change. *Economic Geography, 79*(4), 387–404.

Adger, W. N. (2006). Vulnerability. *Global Environmental Change, 16*(3), 268–281.

Adger, W. N., Kelly, M., Winkels, A., Huy, L. Q., & Locke, C. (2002). Migration, remittances, livelihood trajectories, and social resilience. *Ambio, 31*(5), 358–366.

Adger, W. N., Dessai, S., Goulden, M., Hulme, M., Lorenzoni, I., Nelson, D., et al. (2009). Are there social limits to adaptation to climate change? *Climatic Change, 93*(3–4), 335–354.

Arrow, K. J. (2000). Observations on social capital. In P. Dasgupta, & I. Serageldin (Eds.), *Social capital. A multifaceted perspective* (pp. 3–5). Washington, DC: World Bank.

Arthur, W. B., & Arrow, K. J. (1994). *Increasing returns and path dependence in the economy*. Ann Arbor: University of Michigan Press (Economics, cognition, and society).

Baldacchino, G. (2005a). Editorial: islands—Objects of representation. *Geografiska Annaler B, 87* (4), 247–251.

Baldacchino, G. (2005b). The contribution of 'social capital' to economic growth: Lessons from island jurisdictions. *The Round Table, 94*(378), 31–46.

Baldacchino, G., Kelman, I. (2014). Critiquing the pursuit of island sustainability. Blue and Green, with hardly a colour in between. *Shima: The International Journal of Research into Island Cultures, 8*(2).

Barnett, J., & Adger, W. N. (2003). Climate dangers and atoll countries. *Climatic Change, 61*(3), 321–337.

Bawden, T. (2015). COP21: Support grows for limiting warming to 1.5C amid claims 2C 'is not safe'. In *The independent*. http://www.independent.co.uk/news/world/europe/cop21-support-grows-for-limiting-warming-to-15c-amid-claims-2c-is-not-safe-a6765546.html, checked on January 14, 16.

Betzold, C. (2015). Adapting to climate change in small island developing states. *Climatic Change, 133*(3), 481–489.

Baldacchino, G., & Clark, E. (2013). Guest editorial introduction: Islanding cultural geographies. *Cultural Geographies, 20*(2), 129–134.

Becker, E. (2012). Social-ecological systems as epistemic objects. In M. Glaser, G. Krause, B. M. W. Ratter, & M. Welp (Eds.), *Human-nature interactions in the anthropocene. Potentials of social-ecological systems analysis* (pp. 37–59). New York: Routledge (Routledge studies in environment, culture, and society, 1).

Berkes, F. (2009). Evolution of co-management: Role of knowledge generation, bridging organizations and social learning. *Journal of Environmental Management, 90*(5), 1692–1702.

Berkes, F., & Ross, H. (2013). Community resilience: Toward an integrated approach. *Society & Natural Resources, 26*(1), 5–20.

Berkes, F., Armitage, D. R., & Doubleday, N. (Eds.). (2007). *Adaptive co-management. Collaboration, learning, and multi-level governance.* Vancouver [B.C.]: UBC Press (Sustainability and the environment).

Blaikie, P. M. (1994). *At risk. Natural hazards, people's vulnerability, and disasters.* London: Routledge.

Bohle, H.-G. (2005). Soziales oder unsoziales Kapital? Das Sozialkapital-Konzept in der Geographischen Verwundbarkeitsforschung. *Geographische Zeitschrift, 93*(2), 65–81.

Bourdieu, P. (1983). Ökonomisches Kapital, kulturelles Kapital, soziales Kapital (R. Kreckel, Trans.). In R. Kreckel (Ed.), *Soziale Ungleichheiten* (pp. 183–198). Göttingen: Schwartz (Soziale Welt Sonderband, 2).

Briguglio, L. (2002). The Economic Vulnerability of Small Island Developing States. In: H.-H.M. Hsiao, C.-H. Liu, & H.-M. Tsai (Eds.), *Sustainable development for Island Societies. Taiwan and the World* (pp. 73–89). Taipei: Asia-Pacific Research Program.

Brookfield, H. C. (1990). An approach to islands. In W. Beller, P. d'Ayala, & P. L. Hein (Eds.), *Sustainable development and environmental management of small islands* (pp. 23–34). Paris: UNESCO (Man and the Biosphere Series, 5).

Byrne, J., Inniss, V. (2002). Island sustainability and sustainable development in the context of climate change. In H.-H. M. Hsiao, C.-H. Liu, & H.-M. Tsai (Eds.), *Sustainable development for island societies. Taiwan and the world* (pp. 3–30). Taipei: Asia-Pacific Research Program.

Cambers, G. (2007). Opportunities, not limitations (guest statement). In: G. Baldacchino (Ed.), *A world of islands. An island studies reader.* Charlottetown: Institute of Island Studies, University of Prince Edward Island: 514–515.

Campbell, J. (2009). Islandness. Vulnerability and resilience in Oceania. In *Shima: The International Journal of Research into Island Cultures, 3*(1), 85–97.

Claibourn, M. P., & Martin, P. S. (2007). The third face of social capital: How membership in voluntary associations improves policy accountability. *Political Research Quarterly, 60*(2), 192–201.

Coates, T. (2015). Understanding local community construction through flooding: The 'conscious community' and the possibilities for locally based communal action. *Geo: Geography and Environment, 2*(1), 55–68.

Coleman, J. S. (1988). Social capital in the creation of human capital. *American Journal of Sociology, 94,* 95–120.

Coleman, J. S. (1994). *Foundations of social theory.* Cambridge, Mass., USA: Belknap Press.

Coleman, J. S. (2000). Social capital in the creation of human capital. In: P. Dasgupta, & I. Serageldin (Eds.), *Social capital. A multifaceted perspective* (pp. 13–39). Washington, DC: World Bank.

Crowards, T. (2002). Defining the category of 'small states'. *Journal of International Development, 14,* 143–179.

de Tocqueville, A. (2002) [1835/40]. *Democracy in America. Volumes one and two* (H. Reeve, Trans.). Edited by The Pennsylvania State University (Electronic Classics Series). http://seas3.

elte.hu/coursematerial/LojkoMiklos/Alexis-de-Tocqueville-Democracy-in-America.pdf, checked on May 22, 2012.

Depraetere, C. (2008a). The challenge of nissology: A global outlook on the world archipelago. Part I: Scene Setting the world archipelago. *Island Studies Journal, 3*(1), 3–16.

Depraetere, C. (2008b). The challenge of nissology: A global outlook on the world archipelago. Part II: The global and scientific vocation of nissology. *Island Studies Journal, 3*(1), 17–36.

Diamond, J. (2005). *Collapse*. Viking: How societies choose to fail or succeed. New York.

Dommen, E. (1980). Some distinguishing characteristics of island states. *World Development, 8* (12), 931–943.

Douglas, C. H. (2006). Small island states and territories. Sustainable development issues and strategies—Challenges for changing islands in a changing world. *Sustainable Development, 14*, 75–80.

Durkheim, E. (1933). *The Division of Labor in Society*. New York: The Free Press.

Durkheim, É. (1976). *The elementary forms of the religious life* (2nd ed.). London: Allen and Unwin.

Egner, H. (2008a). Planen, beeinflussen, verändern … Zur Steuerbarkeit autopoietischer Systeme. In H. Egner, B. M. W. Ratter, & R. Dikau (Eds.), *Umwelt als system—System als Umwelt? Systemtheorien auf dem Prüfstand* (pp. 137–154). München: Oekom.

Egner, H. (2008b). Komplexität. Zwischen Emergenz und Reduktion. In H. Egner, B. M. W. Ratter, & R. Dikau (Eds.), *Umwelt als system—System als Umwelt? Systemtheorien auf dem Prüfstand* (pp. 39–54). München: Oekom.

Egner, H., & Ratter, B. M. W. (2008). Einleitung: Wozu Systemtheorie(n). In H. Egner, B. M. W. Ratter, & R. Dikau (Eds.), *Umwelt als system—system als Umwelt? Systemtheorien auf dem Prüfstand* (pp. 9–19). München: Oekom.

Eisenack, K. (2012). Archetypes of adaptation to climate change. In M. Glaser, G. Krause, B. M. W. Ratter, & M. Welp (Eds.), *Human-nature interactions in the anthropocene. Potentials of social-ecological systems analysis* (pp. 107–122). New York: Routledge (Routledge studies in environment, culture, and society, 1).

Field, J. (2008). *Social capital* (2nd ed.). Milton Park, Abingdon, Oxon, New York: Routledge.

Folke, C., Hahn, T., Olsson, P., & Norberg, J. (2005). Adaptive governance of social-ecological systems. *Annual Review of Environment and Resources, 30*(1), 441–473.

Folke, C., Carpenter, S., Walker, B., Scheffer, M., Chapin, T., Rockström, J. (2010). Resilience thinking: Integrating resilience, adaptability and transformability. *Ecology & Society, 15*(4).

Franzen, A., & Pointner, S. (2007). Sozialkapital: Konzeptualisierungen und Messungen. *Kölner Zeitschrift für Soziologie und Sozialpsychologie, 47,* 66–90.

Gaag, van der M. P. J. (2005). *Measurement of individual social capital.* Groningen: University Library Groningen (ICS dissertation series, 115).

Genkin, M. (2006). *A social capital theory of fascism*. The Cases of Germany, Italy, and Japan. Paper presented at the annual meeting of the American Sociological Association, Montreal Convention Center, Montreal, Quebec, Canada, August 10, 2006. http://citation.allacademic.com/meta/p_mla_apa_research_citation/1/0/5/1/8/p105185_index.html, checked on September 04, 2013.

Gillis, J. R. (2012). *The human shore*. Seacoasts in history. Chicago: The University of Chicago Press.

Gillis, J. R. (2014). Not continents in miniature: Islands as ecotones. *Island Studies Journal, 9*(1), 155–166.

Glaser, M., Krause, G., Ratter, B. M. W., Welp, M. (2008). Human/nature interaction in the anthropocene. Potential of Social-Ecological Systems Analysis. *GAIA, 17*(1), 77–80.

Glaser, M., Ratter, B. M. W., Krause, G., Welp, M. (2012). New approaches to the analysis of human-nature relations. In M. Glaser, G. Krause, B. M. W. Ratter, M. Welp (Eds.), *Human-nature interactions in the anthropocene. Potentials of social-ecological systems analysis* (pp. 3–12). New York: Routledge (Routledge studies in environment, culture, and society, 1).

Grootaert, C., & van Bastelaer, T. (2002). *Understanding and measuring social capital. A multidisciplinary tool for practitioners*. Washington, DC: World Bank (Directions in development).

Hallegatte, S. (2012). A framework to investigate the economic growth impact of sea level rise. *Environmental Research Letters, 7*(1), 1–7.

Häuberer, J. (2011). *Social capital theory*. Towards a Methodological Foundation. Wiesbaden: Verlag für Sozialwissenschaften.

Hay, J. E. (2013a). Small island developing states: Coastal systems, global change and sustainability. *Sustainability Science, 8*(3), 309–326.

Hay, P. (2013b). What the sea portends: A reconsideration of contested island tropes. *Island Studies Journal, 8*(2), 209–232.

Hess, A. L. (1990). Overview. Sustainable development and environmental management of small islands. In W. Beller, P. d'Ayala, & P. L. Hein (Eds.), *Sustainable development and environmental management of small islands* (pp. 3–14). Paris: UNESCO (Man and the Biosphere Series, 5).

Hewson, M. (2010). Agency. In A. Mills, G. Durepos, & E. Wiebe (Eds.), *Encyclopedia of case study research* (pp. 13–17). Thousand Oaks, CA: SAGE.

Holdschlag, A., & Ratter, B. M. W. (2013). Multiscale system dynamics of humans and nature in The Bahamas: Perturbation, knowledge, panarchy and resilience. *Sustainability Science, 8*(3), 407–421.

Holling, C. S. (1973). Resilience and stability of ecological systems. *Annual Review of Ecology and Systematics, 4*(1), 1–23.

Hunka, A., & de Groot, W. T. (2011). Participative environmental management and social capital in Poland. *Social Geography, 6*(1), 39–45.

Ireland, P., & Thomalla, F. (2011). The role of collective action in enhancing communities' adaptive capacity to environmental risk: An exploration of two case studies from Asia. *PLoS Currents Disasters, 3,* 1–16.

Jesinghaus, J. (2002). Indicators for decision-making and sustainable development. In H.-H. M. Hsiao, C.-H. Liu, H.-M. Tsai (Eds.), *Sustainable development for island societies. Taiwan and the world* (pp. 289–322). Taipei: Asia-Pacific Research Program.

Jones, N., & Clark, J. (2013). Social capital and climate change mitigation in coastal areas: A review of current debates and identification of future research directions. *Ocean and Coastal Management, 80,* 12–19.

Kelman, I. (2003). The Intrigue of Islands—What are they? *Views (U.K. National Trust), 38*, 70–72.

Kelman, I. (2014). No change from climate change: Vulnerability and small island developing states. *The Geographical Journal, 180*(2), 120–129.

Kelman, I., & Khan, S. (2013). Progressive climate change and disasters: Island perspectives. *Natural Hazards, 69*(1), 1131–1136.

Kerr, S. A. (2005). What is small island sustainable development about? *Ocean and Coastal Management, 48*(7–8), 503–524.

King, R. (1993). The geographical fascination of islands. In D. G. Lockhardt, D. Drakakis Smith, & J. Schembri (Eds.), *The development process in small island states* (pp. 13–37). London, New York: Routledge.

Kithiia, J. (2015). Resourceless victims or resourceful collectives: Addressing the impacts of climate change through social capital in fringing coastal communities. *Ocean and Coastal Management, 106,* 110–117.

Klug, H. (1985). Eine geographische Klassifikation der Inseltypen des Weltmeeres. *Berliner geographische Studien, 16,* 191–217.

Krishna, A., Shrader, E. (1999). Social capital assessment tool. In *Prepared for the conference on social capital and poverty reduction*. Washington: The World Bank. http://siteresources.worldbank.org/INTSOCIALCAPITAL/Resources/Social-Capital-Assessment-Tool–SOCAT-/sciwp22.pdf, checked on July 31, 2013.

LaFlamme, A. G. (1983). The archipelago state as a societal subtype. *Current Anthropology, 24* (3), 361–362.

Lewin, R. (1999). *Complexity*. Life at the edge of Chaos. Chicago: Chicago University Press.

Liebowitz, S. J., & Margolis, S. E. (1995). Path dependence, lock-in, and history. *Journal of Law Economics and Organization, 11*(1), 205–226.

Lin, N. (1999). Building a network theory of social capital. *Connections, 22*(1), 28–51.

Lin, N. (2001). *Social capital. A theory of social structure and action*. Cambridge, UK, New York: Cambridge University Press.

Ling, C., Dale, A. (2014). Agency and social capital. Characteristics and dynamics. *Community Development Journal, 49*(1), 4–20.

Luhmann, N. (1988). Neuere Entwicklungen in der Systemtheorie. *Merkur Deutsche Zeitschrift für europäisches Denken, 42,* 292–300.

MacKinnon, D., & Derickson, K. D. (2013). From resilience to resourcefulness: A critique of resilience policy and activism. *Progress in Human Geography, 37*(2), 253–270.

Maclean, C. (2006). *Island on the edge of the world* (p. 42). The story of St Kilda. Edinburgh: Canongate (Canongate classics.

Magis, K. (2010). Community resilience: An indicator of social sustainability. *Society & Natural Resources, 23*(5), 401–416.

Maida, C. A. (2007). *Sustainability and communities of place*. New York: Berghahn Books (Studies in environmental anthropology and ethnobiology, 5).

Manson, S. M. (2001). Simplifying complexity. A review of complexity theory. *Geoforum, 32*(3), 405–414.

Masten, A. S. (2001). Ordinary magic. Resilience processes in development. *American Psychologist, 56*(3), 227–238.

McCall, G. (1994). Nissology: A proposal for consideration. *Journal of the Pacific Society, 17*(2–3), 1–14.

McElroy, J. (2002). Problems of environmental and social sustainability for small island development. Managing tourism in small islands. In H. -H. M. Hsiao, C. -H. Liu, & H. -M. Tsai (Eds.), *Sustainable development for island societies. Taiwan and the world* (pp. 49–72). Taipei: Asia-Pacific Research Program.

Millennium Ecosystem Assessment. (2005). *Ecosystems and human well-being: Synthesis*. Washington, D.C.: Island Press.

Mohan, G., & Mohan, J. (2002). Placing social capital. *Progress in Human Geography, 26*(2), 191–210.

Moro, D., Agard, J., Vicente, L. (2005). Island systems. In R. M. Hassan, R. Scholes, & N. Ash (Eds.), *Ecosystems and human well-being. Current state and trends, volume 1: findings of the condition and trends working group of the millennium ecosystem assessment* (pp. 663–680). Washington, DC: Island Press (The millennium ecosystem assessment series, v. 1).

Mountz, A. (2015). Political geography II: Islands and archipelagos. *Progress in Human Geography, 39*(5), 636–646.

Nagarajan, P. (2006). Collapse of easter island. Lessons for sustainability of small islands. *Journal of Developing Societies, 22*, 287–301.

Narayan, D., & Pritchett, L. (1999). Cents and sociability: Household income and social capital in rural Tanzania. *Economic Development and Cultural Change, 47*(4), 871–897.

Naughton, L. (2014). Geographical narratives of social capital: Telling different stories about the socio-economy with context, space, place, power and agency. *Progress in Human Geography, 38*(1), 3–21.

Nelson, D. R., Adger, W. N., Brown, K. (2007). Adaptation to environmental change. Contributions of a resilience framework. *Annual Review of Environment and Resources, 32* (1), 395–419.

Nicholls, R. J., Wong, P. P., Burkett, V. R., Codignotto, J. O., Hay, J. E., McLean, R. F. et al. (2007). Coastal systems and low-lying areas. M. L. Parry, O. F. Canziani, J. P. Palutikof, P. J. van der Linden, & C. E. Hanson (Eds.), *Climate change 2007: Impacts, adaptation and vulnerability. Contribution of working group ii to the fourth assessment report of the intergovernmental panel on climate change* (pp. 315–356). Cambridge, UK: Cambridge University Press.

Nunn, P. D. (1994). *Oceanic islands*. Oxford, Cambridge, Mass., USA: Blackwell (The Natural environment, 1).

Nunn, P. D., Aalbersberg, W., Lata, S., & Gwilliam, M. (2014). Beyond the core: Community governance for climate-change adaptation in peripheral parts of pacific island countries. *Regional Environmental Change, 14*(1), 221–235.

Oppenheim, P., & Putnam, H. (1958). Unity of science as a working hypothesis. *Minnesota Studies in Philosophy of Science, 2,* 3–36.

Ostrom, E. (2000). Social capital: A fad or a fundamental concept? In P. Dasgupta, & I. Serageldin (Eds.), *Social capital. A multifaceted perspective* (pp. 172–214). Washington, DC: World Bank.

O'Sullivan, D., Manson, S. M., Messina, J. P., & Crawford, T. W. (2006). Space, place, and complexity science. *Environment and Planning A, 38*(4), 611–617.

Parsons, T. (1951). *The social system*. Glencoe, Illinois: The Free Press.

Pelling, M. (2011). *Climate change and social capital. Foresight Project*. London: Government Office for Science (International Dimensions of Climate Change, 5). http://www.bis.gov.uk/assets/foresight/docs/international-dimensions/11-1029-climate-change-and-social-capital.pdf, checked on November 26, 2012.

Pelling, M., & High, C. (2005). Understanding adaptation: What can social capital offer assessments of adaptive capacity? *Global Environmental Change, 15*(4), 308–319.

Pilkey, O. H., & Young, R. (2009). *The rising sea*. Washington, DC: Island Press/Shearwater Books.

Portes, A. (1998). Social capital. Its Origins and Applications in Modern Sociology. *Annual Review of Sociology, 24*(1), 1–24.

Pugh, J. (2013). Island movements: Thinking with the archipelago. *Island Studies Journal, 8*(1), 9–24.

Putnam, R. D. (1995). Bowling alone: America's declining social capital. *Journal of Democracy, 6* (1), 65–78.

Putnam, R. D. (2000). *Bowling alone. The collapse and revival of American community*. New York: Simon & Schuster.

Putnam, R. D., Leonardi, R., Nanetti, R. (1993). *Making democracy work. Civic traditions in modern Italy*. Princeton, N.J: Princeton University Press.

Radcliffe, S. A. (2004). Geography of development: Development, civil society and inequality—Social capital is (almost) dead? *Progress in Human Geography, 28*(4), 517–527.

Rainbird, P. (2007). *The archaeology of islands*. Cambridge, New York: Cambridge University Press (Topics in contemporary archaeology).

Ratter, B. M. W. (2012). Complexity and emergence. Key concepts in non-linear dynamic systems. In M. Glaser, G. Krause, B. M. W. Ratter, & M. Welp (Eds.), *Human-nature interactions in the anthropocene. Potentials of social-ecological systems analysis* (pp. 90–104). New York: Routledge (Routledge studies in environment, culture, and society, 1).

Ratter, B. M. W. (2013). Surprise and uncertainty—Framing regional geohazards in the theory of complexity. *Humanities, 2*(1), 1–19.

Robinson, G. M., Carson, D. A. (2015). Resilient communities: Transitions, pathways and resourcefulness. *The Geographical Journal*.

Rose, R. (2000). Getting things done in an antimodern society. social capital networks in Russia. In P. Dasgupta, & I. Serageldin (Eds.), *Social capital. A multifaceted perspective* (pp. 147–171). Washington, DC: World Bank.

Royle, S. A. (2001). *A geography of islands. Small island insularity*. London, New York: Routledge.

Royle, S. A. (2007). Definitions & Typologies. In G. Baldacchino (Ed.), *A world of islands. An island studies reader* (pp. 33–56). Charlottetown: Institute of Island Studies, University of Prince Edward Island.

Sawyer, R.K. (2005). *Social emergence. Societies as complex systems* (Vol. 4). Cambridge: Cambridge University Press.

Scheffran, J., Marmer, E., & Sow, P. (2012). Migration as a contribution to resilience and innovation in climate adaptation: Social networks and co-development in Northwest Africa. *Applied Geography, 33,* 119–127.

Schmid, J. (1992). Grundlagen einer modernen Kulturökologie. In: B. Glaeser (Ed.), *Humanökologie und Kulturökologie. Grundlagen, Ansätze, Praxis* (pp. 235–268). Opladen: Westdeutsche Verlag.

Schofield, J., & George, J. J. (1997). Why study islands? In R. A. Irving, C. J. Webster, & A. J. Schofield (Eds.), *Island studies: Fifty years of the lundy field society* (pp. 5–14). Bideford: Lundy Field Society.

Scoones, I. (1998). *Sustainable rural livelihoods: A framework for analysis.* Working Paper 7. Brighton: Institute for Development Studies. http://www.sarpn.org/documents/d0001493/P1833-Sustainable-rural-livelihoods_IDS-paper72.pdf, checked on July 25, 2013.

Scrivens, K., Smith, C. (2013). *Four Interpretations of social capital. An agenda for measurement.* OECD (OECD Statistics Working Papers, 2013/06). http://www.oecd-ilibrary.org/docserver/download/5jzbcx010wmt.pdf?expires=1433321133&id=id&accname=guest&checksum=E4EEA12522A0B73DF5A9C37E436FC7C0, checked on March 06, 2015.

Serageldin, I., Grootaert, C. (2000). Defining social capital: An integrating view. In P. Dasgupta, & I. Serageldin (Eds.), *Social capital. A multifaceted perspective* (pp. 40–58). Washington, DC: World Bank.

Smart, A. (2008). Social capital. *Anthropologica, 50*(2), 409–416.

Smith, J. W., Anderson, D. H., & Moore, R. L. (2012). Social capital, place meanings, and perceived resilience to climate change. *Rural Sociology, 77*(3), 380–407.

Stratford, E., Baldacchino, G., McMahon, E., Farbotko, C., & Harwood, A. (2011). Envisioning the archipelago. *Island Studies Journal, 6*(2), 113–130.

Tanasescu, A., & Smart, A. (2010). The limits of social capital: An examination of immigrants' housing challenges in calgary. *Journal of Sociology & Social Welfare, 37*(4), 97–122.

Teaiwa, T. K. (2007). To Island. Guest statement. In G. Baldacchino (Ed.), *A world of islands. An island studies reader* (p. 514). Charlottetown: Institute of Island Studies, University of Prince Edward Islandp.

Tönnies, F. (2011) [1887]. Soziologische Schriften. Studien zu Gemeinschaft und Gesellschaft. 1st ed. Edited by Klaus Lichtblau. Wiesbaden: VS Verlag für Sozialwissenschaften.

Turner, R. K., Subak, S., & Adger, W. N. (1996). Pressures, trends, and impacts in coastal zones: Interactions between socioeconomic and natural systems. *Environmental Management, 20*(2), 159–173.

United Nations. (1994). Report of the global conference on the sustainable development of small island developing states. In *Global conference on the sustainable development of small island developing states.* Bridgetown, Barbados.

United Nations. (2005). Report of the international meeting to review the implementation of the programme of action for the sustainable development of small island developing states. In *International meeting of the programme of action for sustainable development of small island developing states.* Port Louis, Mauritius: United Nations.

United Nations. (2014a). *Emerging issues for small island developing states.* UNEP: Results of the UNEP foresight process. Nairobi.

United Nations. (2014b). *Report of the third international conference on small island developing states.* Apia, Samoa, September 1–4, 2014. A/CONF.223/10. http://www.un.org/ga/search/view_doc.asp?symbol=A/CONF.223/10&Lang=E, checked on November 01, 2016.

UNEP. (2005). Assessing coastal vulnerability. In *Developing a global index for measuring risk.* Nairobi, Kenya: United Nations Environment Programme.

UN-OHRLLS. (2016). *About the small island developing states.* http://unohrlls.org/about-sids/, checked on January 24, 2016.

UN Secretary General. (2010). Review of the implementation of the mauritius strategy. In *commission on sustainable development. Eighteenth session, May 3–14, 2010, Item 3 of the provisional*. New York. http://www.sidsnet.org/msi_5/docs/sidsday/ECN.17-2010-9.pdf, checked on 19/03/15.

Urry, J. (2005). The complexity turn. *Theory, Culture & Society, 22*(5), 1–14.

U.S. Indian Ocean Tsunami Warning System Program. (2007). *How resilient is your coastal community? A guide for evaluating coastal community resilience to tsunamis and other coastal hazards*. Bangkok: U.S. Indian Ocean Tsunami Warning System Program supported by the United States Agency for International Development and partners. http://www.crc.uri.edu/download/CCRGuide_lowres.pdf, checked on March 04, 2017.

van der Gaag, M., & Snijders, T. A. B. (2005). The resource generator: Social capital quantification with concrete items. *Social Networks, 27*(1), 1–29.

Walker, B.H., & Salt, D. (2006). *Resilience thinking. Sustaining ecosystems and people in a changing world*. Washington, DC: Island Press.

Walker, B., Carpenter, S., Anderies, J., Abel, N., Cumming, G. S., Janssen, M., et al. (2002). Resilience management in social-ecological systems: A working hypothesis for a participatory approach. *Conservation Ecology, 6*(1), 14.

Walker, L. R., & Bellingham, P. (2011). *Island environments in a changing world*. Cambridge: Cambridge University Press.

Walker, B., Holling, C. S., Carpenter, S. R., Kinzig, A. (2004). Resilience, adaptability and transformability in social–ecological systems. *Ecology and Society*, *9*(2).

Weichselgartner, J., & Kelman, I. (2015). Geographies of resilience: Challenges and opportunities of a descriptive concept. *Progress in Human Geography, 39*(3), 249–267.

Wilson, G. A. (2015). Community resilience and social memory. *Environmental Values, 24*(2), 227–257.

Woolcock, M. (2001). The place of social capital in understanding social and economic outcomes. *Isuma: Canadian Journal of Policy Research, 2*(1), 11–17.

Chapter 4
Designing Small Island Research

Having laid out the key concepts, defined terms and identified research gaps, I describe in this chapter the research framework that I apply. Furthermore, this chapter explains underlying assumptions as well as the means and tools for the collection, processing and interpretation of data necessary for answering the research questions.

The framework (see Fig. 4.1) starts with the formulation of specific questions in line with the theoretical approaches, as described in the previous Sect. 3.4. On this basis, I apply two different epistemological approaches. The distinction between the two, however, is not as clear as might seem from the following definitions because the development of the research process was ongoing and the approaches influenced each other. In my case, the hypothesis testing led to expanded research and inclusion of further qualitative methods, which were the basis for hypothesis generation. Qualitative research and the inclusion of elements from the grounded theory approach allowed for consideration of further research perspectives and an extended interpretation of quantitative data beyond the formulated hypotheses.

As discussed in Sect. 3.1, emergence in complex systems implies non-linearity and uncertainty. The epistemological consequence is that 'science must proceed by experiment and observation rather than rational reasoning, since emergent effects are unpredictable before the event' (Sawyer 2005: 32). Therefore, I work with a case study, chosen under certain criteria and subjected to a mixture of hypothesis-generating and hypothesis-testing research.

The research framework lays out the connection of the two different epistemological approaches and the resulting mixed methods approach with several strategic research tools, including a household survey as a quantitative method, combined with expert and stakeholder interviews, participant observation and media/archive analysis as qualitative methods. Finally, evaluation and self-reflection complement the methodology.

The following sections provide a transparent and detailed description of the various elements of my research framework—an important element of scientific work in general, yet an element that is often neglected in the age of mass

J. Petzold, *Social Capital, Resilience and Adaptation on Small Islands*,
Climate Change Management, DOI 10.1007/978-3-319-52225-8_4

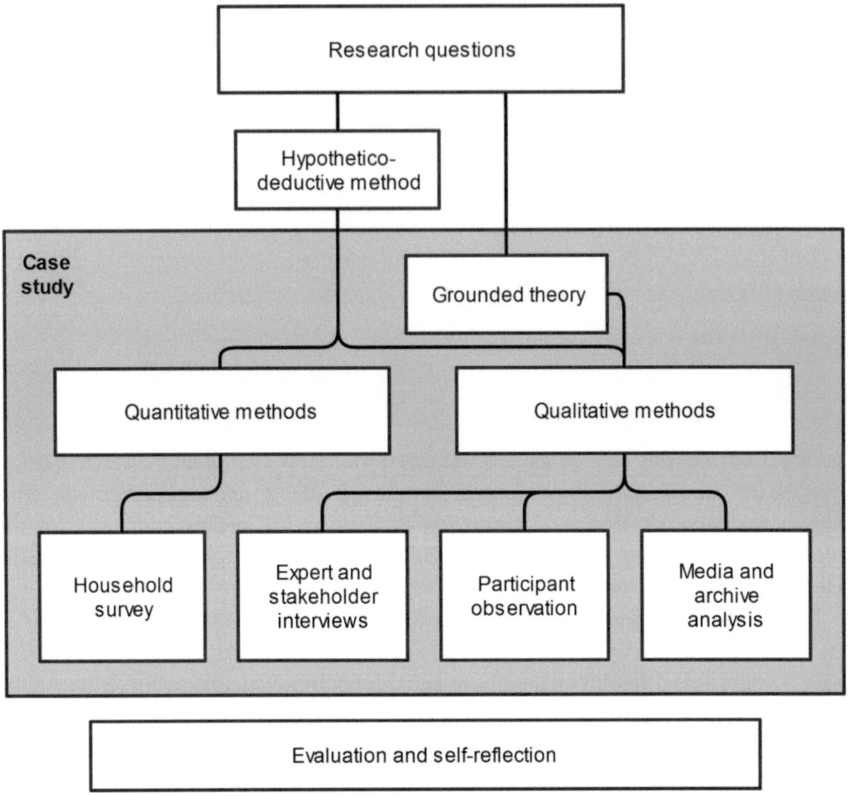

Fig. 4.1 Research framework

publications and interdisciplinary working environments, which climate change research exemplifies (cf. Nielsen and D'haen 2014).

4.1 Epistemological Approach

The key assumption of any scientific endeavour is the understanding of scientific method as 'making incremental improvements in what we know, edging toward the truth but never quite getting there' while keeping the openness 'to have yesterday's truths overturned by today's empirical findings' (Bernard 2006: 4). Central to the modern epistemological approach is to doubt certainties. This approach 'illuminates disciplines from inside and emphasizes how they limit themselves from the start by the presuppositions they accept' (Claval 1980: 373). The problems of positivism and the general uncertainties as discussed, for example, regarding future climate change and its impacts (Chap. 2) certainly also hold true for the social sciences. In fact, 'through focussing analytically on particular themes, patterns or processes in our data, we attempt to infer conclusions about social relationships, processes or causalities that have a broader significance' (Hodkinson 2008: 81).

In contrast to a positivist approach, in which 'external reality awaits our discovery through a series of increasingly good approximations to the truth' (Bernard 2006: 3) and we simply record experiences and translate them into knowledge, constructivism assumes that 'reality is constructed by each person' (Bernard 2006: 3; cf. Berger and Luckmann 1966). On the one hand, '[h]umans continuously engage in material interactions with nature [...]. Thus humans 'construct' their world. This material construction of nature follows historical and cultural patterns as determined by subjective human realities' (Glaser et al. 2008: 77). On the other hand, the experience of 'natural' phenomena depends on social realities. This social construction of nature is accompanied by changing political, cultural and temporal constructs. The epistemological consequence is to deal with 'mental phenomena' (Bernard 2006: 20), not with data from the natural world. It is not a matter of finding 'the' answer, but rather of examining a question and pointing out possible answers to increase understanding (Bernard 2006: 22). I follow this approach to gain an understanding of how social relationships influence people's perception and construction of nature. Moreover, the approach of methodological collectivism for studying social emergence (cf. Sect. 3.1.1) implies a focus on social group practices rather than individual relationships and agency.

I started in a deductive way by applying a hypothetico-dedutive model. At a certain point, however, the increasingly important insights from the direct fieldwork raised further questions, which motivated the adoption of a more inductive approach by integrating elements from grounded theory.

4.1.1 Hypothetico-Deductive Method

The so-called hypothetico-deductive method consists of a general scientific model that works with the formulation of hypotheses and theories to explain, predict and deduce from specific patterns observed. The model is, therefore, 'an alternative to the Baconian "inductive method"' (Jary 2006: 139), which, in contrast, bases theory development purely on observed phenomena. The basis of the hypothetico-deductive model is the formulation of hypotheses, proposition of generalisations and their empirical testing. The results can lead to support or disproof of a theory. Thus, by means of deduction, the model allows for the refinement and advancement of such theories and hypotheses (Jary 2006).

For the combination of inductive and deductive reasoning by testing of ideas against real world data, the development of the hypothetico-deductive model for my research starts with phrasing two major hypotheses, developed on the basis of literature review (Chap. 3) and based on the following axioms [A]:

- [A1] Adaptability towards climate change is increased by facilitated flow of information, public participation and collective action.
- [A2] Collective social capital is a public good and accessible for every member of the community.

The following two hypotheses [H] can be derived:

- [H1] Collective social capital increases adaptability on small islands by facilitating civic engagement and public participation in decision-making processes.
- [H2] Collective social capital increases adaptability on small islands by facilitating flows of information and increasing levels of acceptance towards adaptation measures.

To test these hypotheses, a multi-institutional and multi-scale perspective is necessary. This perspective focuses on two main elements, which are closely inter-related: relationships and networks embodying collective social capital, acting as bridging organisations to governmental institutions (through participation in decision-making, enabling the flow of information from government to people etc.), and those that are enabling individuals to act collectively (apart from connection to the government). In addition, churches/religious communities or other clubs can facilitate the flow and exchange of information but do not stay in contact with institutions in terms of environmental management. However, where are the links (bridging organisations) between environmental management and civil society? Where does social capital play an active role in dealing with coastal protection and adaptation to climate change on small islands?

The analysis starts with the political and institutional context of an island: is the island a dependent territory or an independent island state? Which governing body is responsible for local environmental management, coastal protection and climate change adaptation planning? Secondly, an overview over the organisational context and civil society structure follows: which formal organisations, association, clubs, partnerships or unions are present on the island? Which ones play an active role in environmental management, either by their own activities or by inter-relations with the governmental sphere? For the analysis of the political and organisational context, it is important to keep the perspective over scales, especially the inter-linkages among actors over different scales. This means considering political actors from the municipal level up to national and international political bodies. The same applies to civil society organisations, starting from local grassroots organisations and collective activities up to internationally active environmental non-governmental organisations.

Based on the hypothesis that collective social capital increases adaptability towards climate change impacts on islands, by facilitating decision-making, civic engagement, flows of information and raising acceptance to adaptation measures within the community, I define the following variables and indicators for qualitative and quantitative empirical research (see Fig. 4.2).

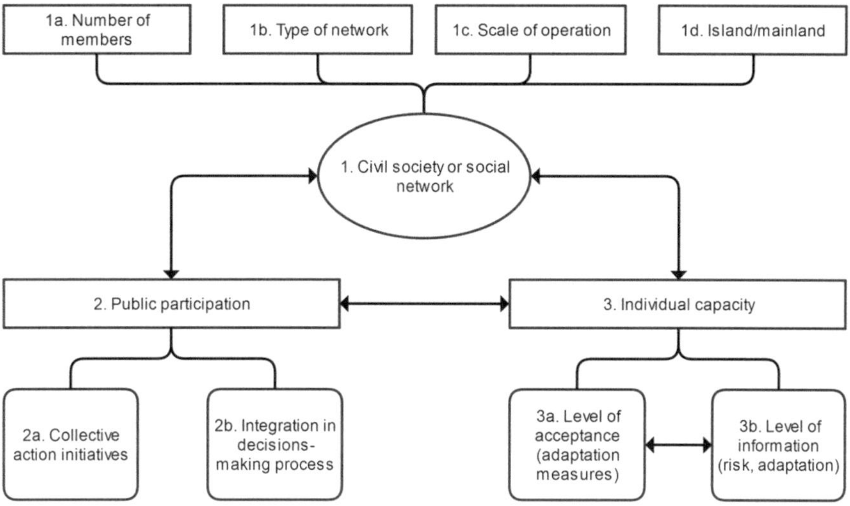

Fig. 4.2 Hypothetico-deductive model

On the side of the independent variable is the network or organisation embodying collective social capital, which needs to be characterised according to the following four features:

- 1a. Number of members
- 1b. Type of network
- 1c. Scale/level of operation (e.g., local/regional)
- 1d. Place of action and origin (island or mainland).

The dependent variables refer to the collective and individual features of social capital and serve as indicators for the mobilisation and impact of social capital:

Public participation:

- 2a. Activity of networks and organisations in terms of collective action involving climate change adaptation
- 2b. Integration of individuals or civil society organisations into processes of political decision-making (co-management).

Individual capacity:

- 3a. Levels of acceptance of adaptation measures within society
- 3b. Levels of information on adaptation measures as well as perception of risks within society.

In contrast to social network analysis, a tool often applied in research related to social capital, the aim here is not to assess personal relationships or central individuals of social networks (cf. Lin 1999: 36), but rather the features that mobilise existing social networks and the areas in which they become relevant to a specific purpose, in this case, climate change adaptation.

To integrate the question of whether the observed effects are *island-specific*, the notion of scale has to be extended from the hierarchical to the geographical level:

- Where do bridging organisations and networks originate and where do they operate?
- Which role do inter-island and island-mainland relationships play, as opposed to intra-island relationships?

The research framework developed through the hypothetico-deductive approach lays the basis especially for the development of the quantitative research methods (Sect. 4.3.1). Due to the nature of the field research, however, aspects of a grounded theory approach found inclusion in the methodology, with consequences especially for the development of the qualitative research methods (Sect. 4.3.2–4.3.4).

4.1.2 Grounded Theory

In contrast to deductive research, grounded theory can be understood as 'empirically based theory formation' (Alheit 1999: 1). A core element of grounded theory is that the research starts 'relatively open and non-prescriptive […] but the imposition of a "preconceived theoretical framework" should be avoided at all costs' (Hodkinson 2008: 85, after Glaser and Strauss 1967: 45). Hypotheses are not formulated at the beginning of the research project, but rather during the research process and repeatedly adjusted. Thus, grounded theory is not only highly open towards the development of the research project, but also iterative, because collected data is being analysed repeatedly with different foci, and hypotheses are developed, discarded and altered throughout the process. Grounded theory is usually applied along with qualitative research methods. An important method that facilitates reflection of data and the researcher's own work is the use of field notes or a research diary (Hodkinson 2008).

Apart from these principles, however, research applying grounded theory can be very diverse. The structure is never as strict as it is in a classical deductive hypothesis testing approach, as described above. Therefore, methodologies of grounded theory studies also differ; e.g., some focus more heavily on the processes of coding data, some less (Hodkinson 2008: 92). Moreover, it is hardly possible to make a complete plan for a grounded theory study at the beginning because it will certainly/naturally change with the progression of the study. Hence, many research projects, including this one, do not strictly follow grounded theory principles but apply certain features, complementary to a previously established research design.

Nonetheless, Hodkinson (2008) outlines steps for a study applying grounded theory. It starts with initial data collection and the first field contact. This phase should be particularly open and not overly focused on specific research questions.

Based on the first analysis of data, which includes data from field notes, more specific research questions and methods can be developed and applied in a focused data collection and analysis. The frequency of this iterative process depends on the individual project and development of hypotheses. Lastly, the theory is refined.

For my research, the concept of grounded theory is important because the hypothetico-deductive model ultimately lacked the openness necessary for developing and reflecting the hypotheses needed to answer my research questions. Bringing together multiple streams of theory in combination with a case study that involves a high degree of involvement of the researcher with his research object is compatible with the principles of grounded theory. Furthermore, grounded theory conforms to the principles of nissology (cf. Sect. 3.3.3).

However, due to the initial development of the hypothetico-deductive research framework, strict adherence to grounded theory principles was not possible. I applied the following elements of grounded theory:

- a very open initial field stay;
- two iterations with refined research questions and refined methods;
- a strong qualitative element, including reflections from systematic field notes.

4.2 Island Case Study

The approach of grounded theory and the qualitative research tools imply an in-depth study of the research object. Therefore, and in complying with the nissological principle to study islands on their own terms, a single case study or a small number of cases makes more sense than large cross-case research. I excluded a comparison between islands from different world regions and an island/mainland comparison so as not to follow constructed dichotomies but rather to focus on specific island cases.

Gerring (2009: 19) defines a case as a 'spatially delimited phenomenon (a unit) observed at a single point in time or over a period of time'. Single case studies, moreover, are important especially for hypothesis generation, in contrast to cross-case studies with a less in-depth qualitative approach, which are adequate for hypotheses testing.

Based on the research background (Chap. 2), theoretical framing (Chap. 3) and available research resources, I developed the following criteria for a suitable case study island:

- Location: European, preferably German- or English-speaking;
- Political status: dependent island territory;
- Low-lying topography;
- Small land area and population, i.e., identifiable issues of scale
- Distance to mainland/peripheral setting resulting in significant issues of location
- Single island or archipelago;

- Climate change-related phenomena in public discussion;
- Data availability (regarding sea-level rise).

The Isles of Scilly match the criteria, as will be described in more detail in Sect. 5.1. As they consist of five inhabited islands and form an archipelagic community, they offer a mixture of a single case study (with a large quantitative survey of the whole archipelago) and cross-case study (by systematically comparing certain variables among the five islands). This distinction is also reflected in general and specific stakeholder interviews. In addition, the case study research partly involved a focus on temporal variations (cf. Gerring 2009: 21).

While islands seem to be suited for relatively bounded case study research, in most cases it is not that simple, as highlighted in Sect. 3.3. Indeed, it is important not to simply lump (in this case five) different islands together and generalise, while at the same time the research object here is one archipelago. Thus, the inter-island connections within the archipelago are important as well as the various intra-island phenomena. To understand these issues, in-depth fieldwork and experience of life on the archipelago are necessary.

The role of fieldwork in social sciences is to 'gain an understanding of the everyday operations and mechanisms of a particular way of life, and the meanings that members of that culture attribute to these everyday occurrences' (Hobbs 2006: 119). Accordingly, the fieldwork included several research stays, complying with the research framework. In fact, the additional field trips resulted from the ongoing development of methods and hypotheses, which are an essential part of the epistemological approach taken.

The first field trip served as an explorative study, including explorative interviews and field notes, identification and confirmation of key stakeholders, setting up contacts and gathering information necessary for the development of the quantitative research tools. The second extended field trip was dedicated to in-depth interviews and conducting the quantitative research. At the same time, the field trip served to adjust the research questions and hypotheses, as well as plan for the integration of further research methods. These were applied in another field trip, with media and archive analysis, as well as additional and more specific interviews. Moreover, the third field trip provided the opportunity to integrate a view of the islands in a different season, which has geographical as well as social implications. I organised a final stay to present the research results to local stakeholders and members of the public, as well as to reflect on the social and political relevance of the research.

4.3 Strategic Methods: Qualitative and Quantitative Research Tools

As already mentioned, the strategic methods applied (i.e., the actual instruments for data collection) consist of a qualitative and a quantitative component. Both approaches influenced each other, and the development of the research tools was accordingly dynamic. Collective action initiatives, participation in clubs and civil society organisations and integration in decision-making are detectable by quantitative methods and, in this case, were partly covered by a household survey. However, these features are regarded as an outcome of social capital, while the actual substance (i.e., the motivation of people to take part, and the promoting and limiting factors) is hardly detectable by such quantitative means (cf. Pelling 2011: 5). Therefore, I applied a mixed-methods qualitative approach, which covered the analysis of the most vulnerable areas on the specific islands and the way the respective island communities collectively deal with risks and impacts of sea-level rise and storm surges. This approach included stakeholder and expert interviews, participant observations as well as literature and media review.

The main period of research included six weeks of field research in February and March 2014 and another three weeks in July 2014. During the first period of field research, the Isles of Scilly experienced the longest and most intensive period of storms in recent decades. Complementary, but important for the actual fieldwork on the islands, was a stay as visiting researcher in Falmouth, Cornwall, which, in addition to providing the advantage of easier organisation of the field work and access to relevant literature, gave me the opportunity to conduct interviews with experts and learn their views of the islands in contrast to the mainland. In addition, the Isles of Scilly's institutional organisation is closely linked with Cornwall and South West England. The following sections describe the detailed preparation and application of the various strategic research methods.

4.3.1 Household Survey

The household survey mainly served to test the hypotheses developed in the hypothetico-deductive model in Sect. 4.1.1. The community size and the time spent on the islands allowed for a distribution to all the households. Therefore, no randomisation or sampling was necessary.

4.3.1.1 Survey Design

The questionnaire titled 'Communities and the sea' was divided into two parts: 'Your community' and 'The sea'. The first part consists of ten questions regarding typical social capital indicators: community membership, trust, participation and information. The second part consists of eleven questions about people's perception of climate change, coastal risks, coastal management and participation in coastal protection. A final part contained statistical questions about place of birth, age and gender. The aim of the survey was to identify bridging networks and organisations and their relevance for the overall adaptability of the islands by connecting the data gained from questionnaires about membership, activities, information and perception directly or indirectly relating to climate change adaptation to the data of the institutional and organisational context. Therefore, data analysis, including correlations among the indicators presented in Sect. 4.1.1 should enable testing of the two hypotheses.

If levels of acceptance, the flow of information, civic engagement and participation are indeed linked to initiatives of collective action, this should be evident from the empirical analysis and demonstrate the role of collective social capital in increasing adaptability on the island. Additionally, bridging organisations can be identified by understanding which elements of social capital are relevant to which of the suggested positive features contributing to the island's adaptability towards climate change. Moreover, an analysis of the relevant organisations identified regarding their origin and operation, as well as sources of information (e.g., on- or off-island), can help explain the role of island-specific features of social capital.

The pilot study was undertaken in two ways. Firstly, I tested the questionnaire with experts in the field of research: two professors at the Department of Geography of the University of Exeter and one from the University of Hamburg. This was followed by a discussion of the scope, comprehensiveness and layout of the questionnaire. Secondly, the questionnaire was tested by a quantitative and qualitative pilot study with a random street population in three different towns in Cornwall (Falmouth, Marazion and Looe), which are all affected by flooding and have similar cultural and societal characteristics to those found on the Isles of Scilly.

For the main household survey on the Isles of Scilly, I chose a drop-and-collect technique. This technique usually produces higher return rates than postal surveys and it makes it possible to reach a high ratio of households in a community, up to all households. The highest return rates are usually achieved with personal distribution as well as collection of questionnaires (cf. Brown 1987; Steele et al. 2001). In my case, however, I made a compromise by providing collection boxes instead of collecting the questionnaires personally. Due to the partially very distant location of certain properties and the distribution of the survey across five islands, this was the only feasible means of distributing the questionnaires to all of the households on the islands in question.

I personally distributed A4 envelopes containing the questionnaire and a cover letter into each residential letterbox on the main island. I placed four collection boxes (Fig. 4.3) in highly frequented and known places in different areas of the island to make it as easy as possible for residents to return their questionnaire. Boxes were placed at the public library and in the one-stop-shop of the local council, both of which are located in the main settlement of the island. Another box was placed in the village shop of the second settlement of the island and a last one was placed in a more rural area at the offices of the local Wildlife Trust.

On the main island, distribution of the questionnaires was met with a few problems. It was difficult to distinguish all-year residential houses from second homes and holiday homes, and there was no complete list of registered residents available. I tried to avoid distributing questionnaires to holiday homes by ringing/knocking at the door or asking neighbours, which sometimes helped, but it was not possible to fully exclude dropping questionnaires at non-resident properties. In case of doubt, I always left one questionnaire in the letterbox.

Fig. 4.3 Survey collection box

I used a slightly different approach on the four smaller islands. On three of them, the distribution was carried out by the local postman and collection boxes were placed at the islands' shops, which are also the post office for each of the islands. On one of the islands, on which touristic facilities and holiday homes are more prevalent, the distribution and collection were carried out by the local estate office, which distributed the questionnaires to their staff, as basically all of the island's residents are employed by the estate. Not distributing the questionnaires on the smaller islands personally was helpful due to the difficulty in distinguishing holiday homes from residential houses, but it was not feasible on the mainland.

The distribution of questionnaires was announced by the local radio, which also broadcasted a reminder after distribution of the questionnaires and placed a short article on their website and social media page. In addition, the social media page of the department of a local environmental organisation called for participation in the survey and I hung posters in several shops.

Residents were given between eight and twelve days to return the questionnaires, depending on the island. The four smaller islands had common drop-off and collection dates. The main island had one collection date, but due to its size, distributing the questionnaires took place over four days. The period for distribution and collection of questionnaires was planned according to local circumstances, e.g., before school holidays started, when many residents would visit the mainland.

4.3.1.2 Survey Outcome

The household survey is most representative for the main island of the Isles of Scilly, St Mary's, where, due the island's size, quantitative figures are more expressive than on the off-islands.

The return rate of the household survey overall is about 21%, with great differences, however, between the various islands (Table 4.1). On St Mary's, almost 20% returned the survey, while on St Agnes, the smallest island community, almost half of the households, and on St Martin's almost 40% returned the survey. On the neighbouring islands Tresco and Bryher the return rate, however, is only about 15%. Overall, the majority of respondents are female (60%; see Fig. 4.4).

Ninety-three per cent of the respondents answered that they spend 10–12 months per year on the Isles of Scilly, which means that the majority of the respondents are local residents and not part of the group of second-home owners, who usually stay on the islands during the summer months. On average, respondents have been living for 32 years on the islands. Only 20% of the respondents, however, were born on the Isles of Scilly (Fig. 4.5). The mean age of respondents is 62, with a minimum value of 23 and maximum of 92. This is reflected also in the age distribution, as shown in Fig. 4.6.

Table 4.1 Turnout of the household survey

	Pilot study	Scilly	St Mary's	St Agnes	St Martin's	Bryher	Tresco
Distributed	17	924	718	31	52	44	79
Collected	17	191	139	14	20	6	12
	100%	20.67%	19.36%	45.16%	38.46%	13.64%	15.19%

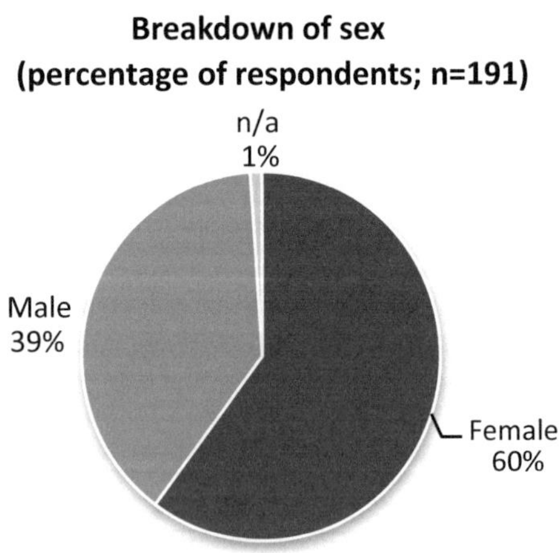

Fig. 4.4 Breakdown of respondents' sex

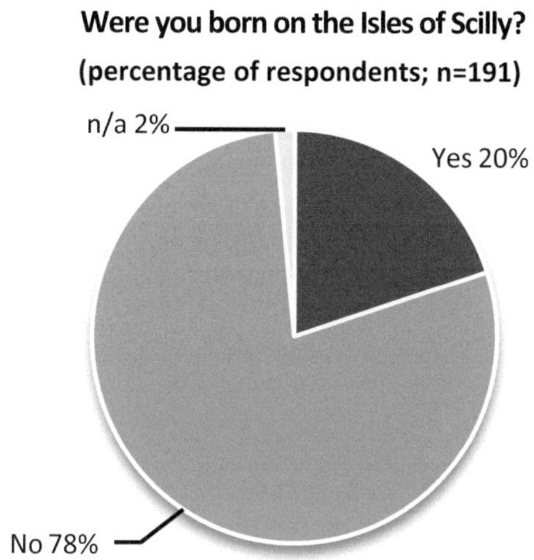

Fig. 4.5 Respondents' place of birth

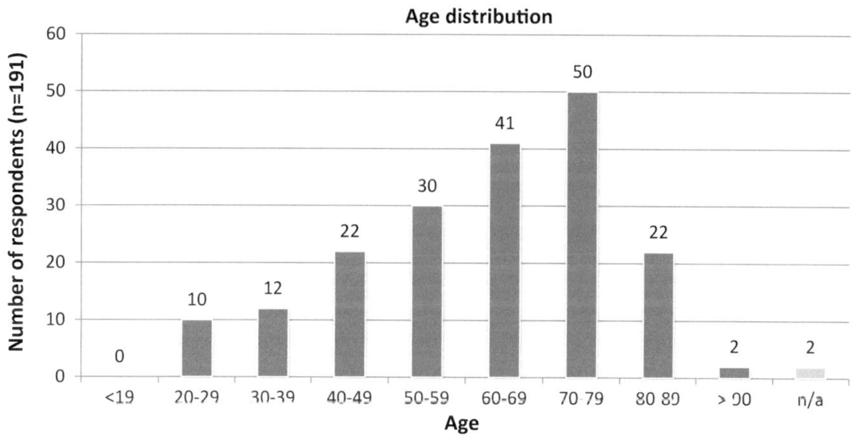

Fig. 4.6 Respondents' age distribution

4.3.2 *Expert and Stakeholder Interviews*

The qualitative interviews consisted of 35 semi-structured interviews with experts, stakeholders and representatives from the community (Table 4.2). The interview partners were chosen by their function in organisations that are relevant to environmental and coastal management, and by the pyramid principle regarding knowledge about social life within their community. Interview guides for the respective interviews included the following central topics:

Table 4.2 Interview partners; codes and total number of interviews for each category in parentheses (interviews will be referred to in the text by the respective code, consecutively numbered, e.g., SC1)

Civil society (SC; 6 interviews)	Administration/government (SP; 18 interviews)	Local population (SL; 8 interviews)	Media (SM; 3 interviews)
• Representatives from environmental organisations • Representatives from collective action initiatives	• Representative from landholder • Officers from local authority • Councillors from each island	• Local residents with family roots on island • Local residents initiating collective action	• Local radio • Freelance journalist • Editor of local magazine

- the vulnerability of the islands to climate change;
- adaptation measures (past and present);
- key stakeholders in decision-making;
- participation of the population in decision-making and planning;
- participation of the population in coastal protection;
- initiatives of collective action within the communities;
- social cohesion: levels of trust and cooperation on islands;
- the role of clubs, associations and civil society;
- levels and sources of information for the public;
- historical change of the community and consequences for the mobilisation of social capital;
- future perspectives for the island community.

Most of the interviews were one-to-one interviews, which I recorded and transcribed (anonymised). I used the software MAXQDA for coding the transcripts, initially based on the categories (with sub-categories) of the interview guides. The coding and interpretation of transcripts were an iterative process, in line with the approach of grounded theory. The various field trips and respective phases of interviews each contributed to the development of the subsequent phase, as well as to the interpretation of data from the household survey.

4.3.3 Participant Observation

The second qualitative method I applied was participant observation. This helped me to understand everyday life within the community and the potential of the community to respond to flood and storm impacts. Participant observation proved particularly relevant to this study, due to unforeseeable extreme weather events during the first period of research. In this way, I was able to directly observe the most vulnerable areas, storm impacts and the reaction of local authorities and the community to the events.

Table 4.3 Attendance of meetings and events

Official meetings and community events attended (codes in parentheses)	Date/place
Meeting of the South West Regional Flood and Coastal Committee	16/01/2014, Exeter
Joint Advisory Committee of the Area of Outstanding Natural Beauty (AONB)	11/12/2014, Isles of Scilly
Self-organised beach cleaning by locals	08/03/2014, Isles of Scilly
Weekly gig rowing race	16/07/2014, Isles of Scilly
Public discussions session with Andrew George (locally elected MP)	11/07/2014, Isles of Scilly

One component of this work was the systematic attendance of official meetings and events (Table 4.3). Observations here included questions on participation (individual actors and demographic characteristics): the number of participants, issues raised and discussed and by whom these issues were raised. I particularly focused on the connection between actors from different social or organisational backgrounds. Like the expert and stakeholder interviews, I coded and analysed the notes from the participant observations using the software MAXQDA.

In addition to the events listed in Table 4.3, I had the opportunity to become acquainted with local residents and their everyday lives during my two extended research stays in February/March and July 2014. Living on the islands for more than two months gave me the opportunity to gain an insight into community life, relevant clubs and organisations, key figures for community activities and collective action, potential areas of conflict, the work of the council and everyday life problems and disruptions caused by storms and flooding. I systematically recorded these as field notes in a daily research dairy. Hobbs (2006: 120) defines field notes as 'selective accounts which, when accumulated over a period of time, will form the core of an academically framed account of a culture'. As such, this method enables 'cultural interpretation to take place' (Hobbs 2006: 120). I coded and analysed the field notes using MAXQDA.

To supplement the observations recorded in my field notes, I made photographs of observed issues including the development of coastal erosion, the condition of coastal defences and the character of community events.

4.3.4 Media and Archive Analysis

I applied literature and media review as a complementary approach to each of the sections of interest: past and present vulnerability to climate change impacts, the perception of risks, the development of the community and the role of social capital in the adaptation to climate change. Key documents for the literature analysis included the recent economic strategy report by the consultancy Ash Futures

(2014), the Climate Change Strategy of the Council of the Isles of Scilly (2010), the AONB Unit's Management Strategy (2010), the report for English Heritage by Johns, Larn et al. (2004) and additional historically relevant publications by Matthews (1960), Ashbee (1974), Bowley (1980) and Thomas (1985).

As there is no local newspaper on the Isles of Scilly, I analysed the most important local media sources (magazines):

- *Scilly Up to Date* (1985–2001);
- *Scilly Now & Then* (2003–2014);
- *The Scillonian* (1925–2014).

In addition, I included news articles from the currently most important source of information on the islands, *RadioScilly* (2011–2015). These articles are archived online at *ScillyToday* (www.scillytoday.com) dating back to 2011. From all the above sources, I considered articles that included the following keywords:

- storms;
- floods;
- erosion;
- climate change;
- sea-level rise;
- sea defences;
- collective action.

In addition to using articles as sources of information, I analysed the media sources and the photo archive of the Isles of Scilly Museum for historical photographs related to the keywords above. The photographs facilitated an understanding of land use change, especially in the coastal zone, impacts of extreme events and collective action initiatives.

4.3.5 Evaluation and Self-reflection

Finally, another method used was the ongoing reflection of my research process, which included:

- an evaluation of the research methods applied during the fieldwork;
- the refinement of hypotheses and questions;
- and the influence of feelings and emotions on my research.

The major tool for this method was the research diary for recording field notes, which has already been mentioned.

I list it as a separate research method because of its high relevance to my research. It helped me to question the validity of research methods and approaches and resulted in the integration of additional methods. At the same time, the reflection on personal influences on the research helped me, e.g., to critically interpret face-to-face interviews, my own perception of people's behaviour and the

Fig. 4.7 Presentation of research results at the Isles of Scilly Museum, 08/09/2015

impacts of environmental phenomena I observed. Moreover, it is the foundation of grounded theory and complies with the nissological principle of *depth*, i.e., remaining open-minded towards results, even with regard to basic questions.

In line with my approach to reflect upon my research, and related to the nissological principles of *height* and *width* (i.e., to communicate research outcomes to decision-makers as well as society), I organised a public presentation and discussion session during my final field trip to the islands (Fig. 4.7).

References

Alheit, P. (1999). Grounded Theory: Ein alternativer methodologischer Rahmen für qualitative Forschungsprozesse. http://www.fallarchiv.uni-kassel.de/wp-content/uploads/2010/07/alheit_grounded_theory_ofas.pdf. Checked on 19/06/15.

Ash Futures. (2014). *Island futures. A thriving, vibrant community rooted in nature, ready for change and excited about the future*. A strategic economic plan for the Isles of Scilly. Ashburton. http://www.scilly.gov.uk/sites/default/files/events/Island%20Futures%20final.pdf. Checked on 08/05/14.

Ashbee, P. (1974). *Ancient Scilly from the first farmers to the early Christians. An introduction and survey*. Newton Abbot: David & Charles.

Berger, P. L., & Luckmann, T. (1966). *The Social Construction of Reality*. A Treatise in the Sociology of Knowledge. New York: Penguin.

Bernard, H. R. (2006). *Research methods in anthropology. Qualitative and quantitative approaches* (4th ed.). Lanham, MD: AltaMira Press.

Bowley, R. L. (1980). *The fortunate islands. The Story of the Isles of Scilly* (7th ed.). St. Mary: Bowley.

Brown, S. (1987). Drop and collect surveys: A neglected research technique? *Marketing Intelligence & Planning, 5*(1), 19–23.

Council of the Isles of Scilly. (2010). *Climate change strategy 2010.* Draft 1. http://www.scilly.gov.uk/environment/Climatechange.htm. Checked on 12/04/13.

Claval, P. (1980). Epistemology and the history of geographical thought. *Progress in Human Geography, 4*(3), 371–384.

Glaser, B.G. and Strauss, A.L. (1967). *The Discovery of Grounded Theory. Strategies for Qualitative Research.* New Brunswick and London: Aldine.

Glaser, M., Krause, G., Ratter, B. M. W., & Welp, M. (2008). Human/Nature Interaction in the Anthropocene. Potential of Social-Ecological Systems Analysis. In: GAIA *17*(1), pp. 77–80.

Gerring, J. (2009). Case study research. Principles and practices. Reprinted. Cambridge: Cambridge University Press.

Hobbs, D. (2006). Fieldwork. In V. Jupp (Ed.), *The SAGE dictionary of social research methods* (pp. 119–121). London: SAGE.

Hodkinson, P. (2008). Grounded theory and inductive research. In G. N. Gilbert (Ed.), *Researching social life* (3rd ed., pp. 80–100). Los Angeles: SAGE.

Isles of Scilly AONB Unit. (2010). *Isles of Scilly. Area of outstanding natural beauty.* Caring for Scilly's exceptional landscape, supporting sustainable island communities. AONB Management Strategy 2010–2014. Hugh Town. http://www.ios-aonb.info/wp-content/uploads/2012/09/IoS-AONB-STRATEGIC-PLAN-2010-2014-LR.pdf. Checked on 12/04/13.

Jary, D. (2006). Hypothetico-deductive model. In V. Jupp (Ed.), *The SAGE dictionary of social research methods* (pp. 138–140). London: SAGE.

Johns, C., Larn, R., & Tapper, B. P. (2004). *Rapid coastal zone assessment for the Isles of Scilly.* A Report for English Heritage. Report No: 2004R030. Historic Environment Service. Truro. http://www.english-heritage.org.uk/publications/isles-of-scilly-rczas/islesofscilly20080116095450.pdf. Checked on 20/08/14.

Lin, N. (1999). Building a network theory of social capital. *Connections, 22*(1), 28–51.

Matthews, G. F. (1960). *The Isles of Scilly. A constitutional, economic and social survey of the development of an island people from early times to 1900.* London: George Ronald.

Nielsen, J. Ø., & D'haen, S. A. L. (2014). Asking about climate change: Reflections on methodology in qualitative climate change research published in Global Environmental Change since 2000. *Global Environmental Change, 24,* 402–409.

Pelling, M. (2011). Climate change and social capital. Foresight project. Government office for science. London (international dimensions of climate change, 5). http://www.bis.gov.uk/assets/foresight/docs/international-dimensions/11-1029-climate-change-and-social-capital.pdf. Checked on 26/11/12.

Sawyer, R. K. (2005). *Social emergence: Societies as complex systems.* 4. pr. Cambridge: Cambridge University Press.

Steele, J., Bourke, L., Luloff, A. E., Liao, P.-S., Theodori, G. L., & Krannich, R. S. (2001). The Drop-off/pick-up method for household survey research. *Journal of the Community Development Society, 32*(2), 238–250.

Thomas, C. (1985). *Exploration of a drowned landscape. Archaeology and history of the Isles of Scilly.* London: B.T. Batsford.

Chapter 5
Climate Change and Community Action on the Isles of Scilly

Nowhere in the Isles of Scilly can one forget the sea. For the sea both unites and divides the islands. (Ashbee 1974: 38–39)

The sea is both a threat and an opportunity, […] a double-edged sword […] for us. (SP10: 16)

As a group of small islands confronted with risks associated with climate change, the Isles of Scilly are a highly interesting case study for multiple reasons: the islands have experienced constant change in terms of political, social and economic circumstances, ranging from periods of flourishing to periods of starvation. Today, their beautiful land- and seascapes attract loyal visitors who return from year to year and cherish the islands. Yet the future of the islands is now uncertain—not least because of the threat of climate change on the horizon.

According to the consultancy Royal Haskoning, which drafted the Shoreline Management Plan (SMP) for Cornwall and the Isles of Scilly, 'one of Scilly's greatest assets' is the 'extremely strong sense of community and […] cohesion and sense of identity within the local population' (Royal Haskoning 2010: 1). Yet questions about how this social cohesion manifests itself in the community and how people perceive and react to climate change impacts still need to be answered.

The combination of research methods I applied produced a wide range of empirical data, both quantitative and qualitative, and provided the basis for answering these questions through a thorough analysis of the islands' social capital and capacity for adaptation to climate change.

5.1 Case Study: The Isles of Scilly

The first section of this chapter gives an overview of the islands' natural environment and potential local impacts of climate change. In particular, I focus on sea-level rise and observed storm and flood conditions, which already affect the islands. The second section focuses on the islands' historical development up to today, in terms of economic, political and, especially, social structures and changes.

J. Petzold, *Social Capital, Resilience and Adaptation on Small Islands*,
Climate Change Management, DOI 10.1007/978-3-319-52225-8_5

Each section, moreover, contains a part that stresses relevant environmental and social differences between the various islands that constitute the Isles of Scilly.

5.1.1 *Environment and Vulnerability of the Isles of Scilly*

The Isles of Scilly are an archipelago 45 km off the southwest coast of the United Kingdom. Though they are often considered part of Cornwall, '[t]his geographical distance [...] and the significant depth of water in between [...] dictates that there are no physical process links with the mainland' (Royal Haskoning 2010: 7). The mean spring tidal range at St Mary's is 4.9 m and the islands are shaped by the 'extremely influential nature of the wave climate' (Royal Haskoning 2010: 8). Composed of the five inhabited islands St Mary's, Tresco, St Martin's, Bryher and St Agnes, as well as many more islets and rocks, mainly consisting of granite and sand accumulations (Royal Haskoning 2010: 8), each island of the Isles of Scilly faces particular impacts of sea-level rise and erosion and responds in its own way. The isles can be counted among the most vulnerable island regions in Europe, for they are low-lying, directly affected by sea-level rise and increasing westerly winter storms, in addition to being peripheral and isolated. Thus, 'over the next 50 to 100 years, there is a risk that the islands maybe become uninhabitable, because of the increased sea-level rise and also the increased impact of storms' (SP1: 6); 'the islands are hugely vulnerable to [...] climate change' (SP10: 20). In its climate change strategy, the Council of the Isles of Scilly states that '[i]t is clear that the biggest threat to the Islands will come from rising sea levels coupled with a combination of exacerbating factors such as storm surge, wind and weather conditions' (Council of the Isles of Scilly 2010: 29).

Other possible impacts of climate change relevant to the islands, especially due to changing seasonal patterns and warming atmospheric temperatures, include changes affecting local bird populations and migratory birds coming to the islands (SC5: 17; SM2: 111), changing fish stocks, effects on the growth seasons of locally cultivated flowers (SP6: 46) and water availability during dry summers (SP10: 54). As all these sectors are vital to the islands' economy, a representative from the council summarises that 'if the climate does change significantly here, then we are going to have a problem' (SM1: 29). The most obvious and immediate impacts of climate change, however, are sea-level rise and increases in storm frequency and intensity.

The archipelago has a long history of submergence. The highest point on the Isles of Scilly, Telegraph Hill on St Mary's, is 49 metres above sea level; in fact, most of the islands comprising the Isles of Scilly today were the hilltops of a single, larger island, still partly connected by the intertidal zone (Fig. 5.1) (Mortimer et al. 2013). Sea levels have risen continually in South West England and on the Isles of Scilly over the last century (Araújo and Pugh 2008) (see Figs. 5.2 and 5.3), in recent years at a rate of approximately 3 mm per year (SP1: 18). Indicators of submergence (erosion within the last decades and century, as well as archaeological

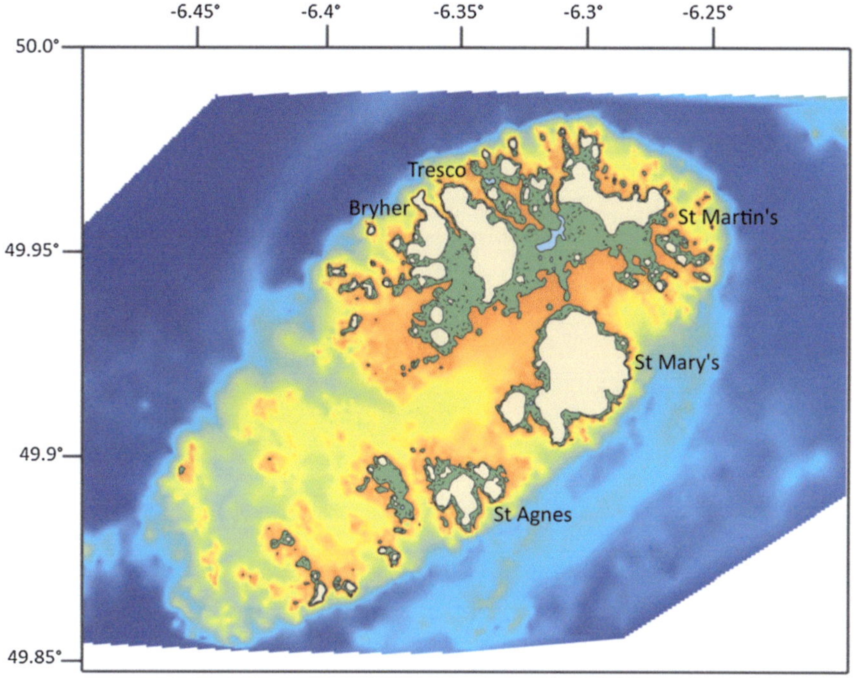

Fig. 5.1 Map of the Isles of Scilly showing the present extent of the inter-tidal zone (*green*), land area (*brown*), and bathymetry from shallow (*yellow*) to deep (*blue*) (*Source* Mortimer et al. 2013: 156)

findings, submerged fields and artefacts) reveal the longer history of sea-level rise on the islands and give the impression of Scilly as a 'drowned landscape' (Thomas 1985). As early as 1822, the *Review of the present state of the Scilly Islands* mentions the risk of further submergence and separation of the islands:

> [U]nless some efficient methods be speedily adopted, the next generation may contemplate [...] the formation of more islets by separation of the land; houses buried in the sand; wand water flowing, in confessed and indisputable sovereignty, over the subjected shore. (Woodley 1822: 244).

As commented in *The Scillonian* magazine, 'parts [...] will be submarine...one day...whatever measures are taken. Scillonians knew that well before hydrogeology became a science' (Mumford 2001: 2).

The ongoing submergence of the islands has drowned ancient settlements and agricultural areas, which can still be observed today at low water levels and by diving (cf. Johns et al. 2004; field notes 13/07/2014). Processes of erosion reveal archaeological artefacts and threaten important historical sites that are not yet submerged. The Isles of Scilly are a highly protected area, designated an Area of Outstanding National Beauty (AONB), a European Marine Site (EMS) as well as home to 26 Sites of Special Scientific Interests (SSSI). The islands feature 238

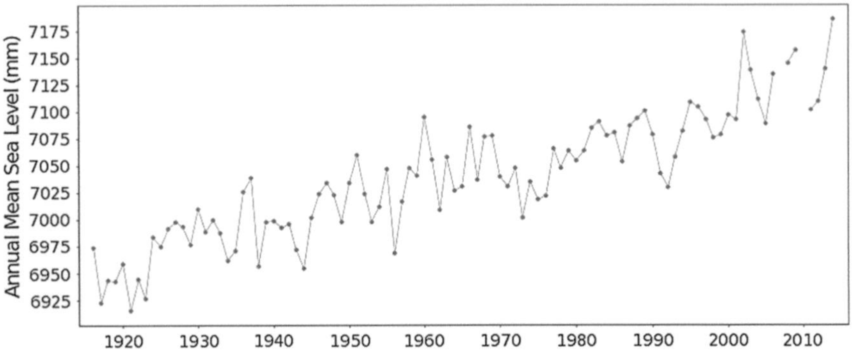

Fig. 5.2 Annual mean sea levels since about 1920 in Newlyn (*Source* Permanent Service for Mean Sea Level 2016b)

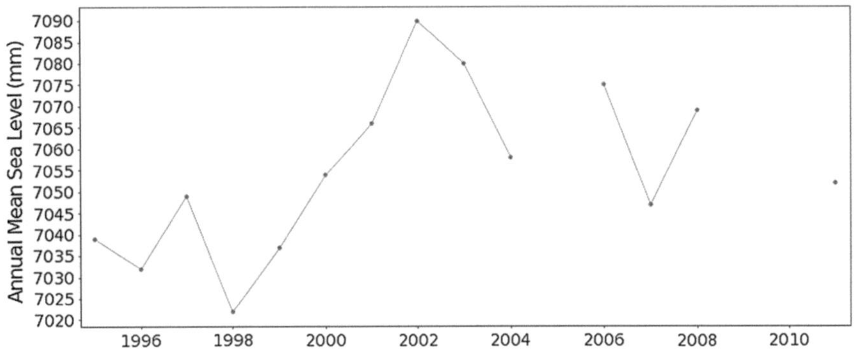

Fig. 5.3 Annual mean sea levels since 1995 in St Mary's (*Source* Permanent Service for Mean Sea Level 2016a)

Scheduled Monuments of archaeological relevance (Isles of Scilly AONB Unit 2010) and at the same time the 'highest number of Scheduled Monuments at risk from coastal erosion in the entire South West Region' (Royal Haskoning 2010: 4) (Fig. 5.4).

Literature and media analysis showed that storms have always affected life on the Isles of Scilly (see Fig. 5.5). Storms and the submerging landscape are part of the Scillonian identity. Storm activity and impacts on the islands feature prominently in all of the media sources analysed, throughout their period of publication. Several publications (Woodley 1822; Heyworth and Kidson 1982; Thomas 1985; Johns et al. 2004) and contributions to magazines and newspapers (e.g. Sisam 1962; Mumford 1985a, b; Tregarthen 2000; Mumford 2001; Tangye 2007; Smith 2008; Mumford 2010) also deal with the issue of sea-level rise on the islands. The high number of ship accidents due to the rough seas around the islands is also a recurring theme in local media accounts; many shipwrecks can still be found near the Isles of Scilly (cf. Larn 1994).

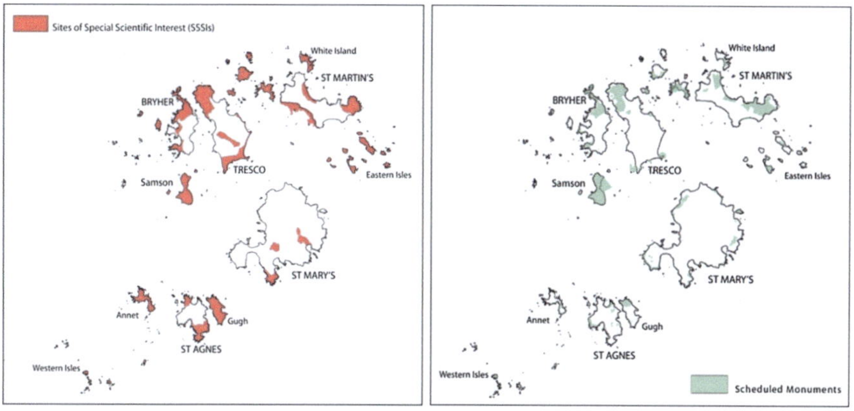

Fig. 5.4 Distribution of SSSIs and Scheduled Monuments on the Isles of Scilly (*Source* Isles of Scilly AONB Unit 2010: 38, 63)

Fig. 5.5 Surge hitting Old Quay, St Mary's, in 1998 (*Source* both, Read 1998b)

Today, rising sea levels, erosion, high tides and storm surges are a major concern for the local population and decision-makers regarding climate change. Future projections for South West England suggest a further sea-level rise of about 20 cm by 2040 and 50 cm by 2100, increased by processes of glacial isostatic adjustment (GIA) (Council of the Isles of Scilly 2010). Grinsted et al. (2015), based on a summary of the latest median projections for twenty-first century sea-level rise in Northern Europe, project a rise of as much as 80 cm. In addition to further erosion and submergence, sea-level rise of this magnitude would lead to serious flooding of infrastructure and housing, not only during peak tides (SP1: 74; SP10: 20), as well as contamination of agricultural areas and freshwater sources (SC2: 18).

Before and during my first field trip there in February/March 2014, the Isles of Scilly experienced one of the longest and most intensive period of storms in recent decades (Figs. 5.6 and 5.7). Therefore, particular emphasis in the following section is placed on the recent storm impacts.

Fig. 5.6 Storm surge hitting Porthcressa on February 14, 2014 (*left*), and overtopping sea wall (*right*) (*Source* both, A. Martin, private photograph, 14/02/2014)

Fig. 5.7 Damaged sea wall and slipway at Porthcressa, February 15, 2014 (*Source* both, A. Martin, private photograph, 15/02/2014)

The effects of storms are felt in everyday life on the islands in various ways. Storms and storm surges generally hinder any outdoor labour, cause power cuts and, especially, disruptions to transport links to the mainland as well as between the islands (SC1: 6; Mumford 1998: 9–10, 1999, 2005). During the stormiest winter months, there is no ferry connection to the mainland. The only way to reach or leave the islands is by plane, which is operated by the Isles of Scilly Steamship Company. There are several flights daily, except on Sundays, to airports in Cornwall (Newquay and Land's End) and Exeter. The connection, however, is very unreliable, because the small aircrafts cannot take off or land during stormy or foggy weather. Therefore, delays or cancellations are common in the winter (SP13: 14) and perceived as normal by many islanders (field notes, 05/02/2014, 14/03/2014). Slightly greater inconveniences are caused by very rough weather conditions, when the freighter, which provides food and other supplies to the islands three times a week, cannot sail. The only supermarket on the Isles of Scilly, on St Mary's, and the small shops on the other islands rely on the freight ship. As such, islanders are confronted with empty shelves in the stores when the ship cannot

Fig. 5.8 Empty supermarket shelves in Hugh Town, 16 February 2014

sail (Fig. 5.8). The closure of roads due to flooding, provision and filling of sandbags and installation of floodgates are ordinary actions and issues for islanders in stormy weather conditions. Usually, there is little surface water flooding due to the quick runoff and tidal cycles (SP1: 14).

Sea-level rise and storms affect the individual islands differently. There are vulnerable areas on each of the inhabited islands, with different consequences for sea defences, local planning and challenges for the local communities. The following sections describe specific vulnerabilities on the islands in a historical context and at present, based on archive and media analysis, expert interviews and especially my own observations during the strong storm period in the winter of 2013/2014.

5.1.1.1 St Mary's

Although housing on the Isles of Scilly is generally not located in the vulnerable coastal areas, Hugh Town, St Mary's main settlement, is located on an isthmus and at sea level (Fig. 5.9). This can be explained by the nearby garrison, which provided protection for the community in former times, when shipbuilding increased the island's wealth and people began to construct more houses (SP3: 48). There are historical records of multiple severe storm impacts and erosion in Hugh Town, such as in 1744, 1820, 1850, 1924, 1962 (Read 1990a), the 'most violent gales in living memory' in 1974 (Mumford 1974: 12) and the 'worst storm of the century' (Read 1990b: 19) in 1989. The Isles of Scilly Museum archived a quote of the garrison officer Robert Heath after the storms in 1744: 'The Town, as it is at present situated, is subject to be destroyed by the inundation of the sea' (Read 1990a: 11).

Today, the town is home to about two-thirds of the island's population and still the most vulnerable area on the Isles of Scilly to sea-level rise and coastal flooding.

Fig. 5.9 Flooded pubs, 14 February 2014 (*left*) and November 1984 (*right*, *source* Gibson 1985: 7)

The main facilities, shops, hotels, restaurant/pubs, as well as housing and the quay often suffer damages and flooding from strong gales and storm surges (cf. Mumford 1990) (Fig. 5.9). On one side of the sand bar, the town is protected by a sea wall at Holgates Green and by the quay, both of which, however, suffer from storm damage; the quay also has to be closed during the most extreme weather conditions (cf. Mumford 1974). On the other side, Porthcressa, hard sea defences protect the town, which could not prevent damage and flooding of properties during recent storms (see Fig. 5.10) (SP7: 2–5).

Outside Hugh Town, several areas are vulnerable to flooding from storm surges and coastal erosion. The greatest impact of erosion can be observed near the peninsula Peninnis, where footpaths have been repeatedly rerouted and gardens and allotments are at risk of falling away (SM1: 43). Behind Porthmellon lies the industrial estate, which is partly below sea level. If the dunes at Porthmellon were to be breached, then this area would flood, including the fire station, local galleries and workshops, the brewery, a development office, the radio station and other facilities.

Fig. 5.10 Hugh Town on St Mary's (*left*, December 2013; *right*, September 2015)

During the heavy storm period in the winter of 2013/2014, the dunes were not breached, but the road just between the industrial estate and the beach was partly flooded by water surpassing one of the floodgates. At Porthlow, dunes were heavily eroded, leaving nearby housing, workshops and boatyards highly vulnerable to flooding. At Lower Moor, a low-lying swamp area spanning from the industrial estate at Porthmellon to Old Town Bay on the other side of the island, the first inundations due to sea-level rise are expected to take place (SP4: 4) (see Appendix A.1). In the long term, this would imply a split of the islands into two parts (SM1: 53; SC3: 33).

Throughout the island of St Mary's, constant erosion is affecting archaeological sites and footpaths, historical sites such as the garrison walls, various allotments and small-scale agricultural areas (Fig. 5.11; field notes, 09/03/2014; SP3: 140). Despite protection from high hedges, strong gales can cause 'widespread damage' (Mumford 2008a) of the flowers cultivated on the islands (cf. Mumford 1974, 1998).

The second-largest settlement, Old Town, is also affected by flooding (Fig. 5.12). The road that lies just behind Old Town Bay, connecting Hugh Town to Old Town and the airport, has to be closed during flood events, which also affects access to the local school. School children then have to be escorted home by emergency services (field notes, 07/02/2014). Whether these bays and settlements

Fig. 5.11 Erosion around St Mary's (*top* and *lower left* December 2013; *top* and *lower right*, September 2015)

Fig. 5.12 Flooding in Old Town, 2005 (*Source* both, Mumford 2005: 19, 21)

are affected by flooding and storms depends primarily on high tides combined with surges and, especially, on wind directions. An unusual combination of these factors caused the impacts of the storms on the Isles of Scilly in the winter of 2013/2014.

5.1.1.2 Off-Islands

The so-called 'off-islands' include the four smaller inhabited islands, Tresco, St Martin's, St Agnes and Bryher, located between 2 and 4 km off St Mary's.

Under predominant (south-)westerly wind directions, the largest off-island, Tresco, is rather sheltered because it is protected by its neighbouring island Bryher (see Appendix A.2). Storms, therefore, usually do not greatly affect Tresco (cf. SP3: 30; SL8: 121). Nonetheless, the island is vulnerable to rising sea levels, because its main settlements, New Grimsby and Old Grimsby, as well as Tresco Abbey and its famous botanical gardens (one of the major tourist attractions on the Isles of Scilly) are very low-lying (SL4: 6). The southern part of Tresco, mainly consisting of large sand dunes and beaches, suffered from heavy erosion, losing several metres of land area during the winter storms in 2013/2014—altogether around 12 metres in the past three years alone (SP18: 2). Those events also exposed the electricity cable from St Mary's to Tresco. One property in Old Grimsby and some farmland were flooded. Sandbanks in the shallow Tresco Channel between Tresco and Bryher moved significantly, forcing local boat services to adjust their routes to the other islands (field notes, 28/07/2014).

On Bryher, the most vulnerable areas regarding sea-level rise and flooding are the southern and southwestern parts of the island (see Appendix A.2), which are exposed to the main wind directions. Strong winds combined with high tides cause overtopping of the sea defences and flooding of the pool area (Fig. 5.13). Some of the dykes suffered from heavy erosion during the last storms and required repair. Some uninhabited land, including part of a Site of Special Scientific Interest (SSSI), was flooded. The low-lying southern part of Bryher could become detached from

Fig. 5.13 Seawall on Bryher built in the 1960s (*left*, February 2014), overtopped by a surge in 2014 (*right*, *source* R. Pearce, private photograph, February 2014)

Fig. 5.14 Erosion on St Agnes (*left*; February 2014); closed footpath due to erosion (*right*; September 2015)

the rest of the island if dunes and seawalls are further eroded and overtopped by increased storm activity and higher sea levels (SP15: 84; Mumford 1993).

The island of St Agnes suffers from ongoing erosion (Fig. 5.14 and Appendix A.3) and was severely affected by the storms in the winter of 2013/2014, damaging critical sea defences on the western part of the islands, known as Periglis (cf. Mumford 2000). These sea defences protect an SSSI, fresh water sources, multiple tourist properties and an electricity sub-station. In addition, the area around the island campsite was affected by erosion and several metres of coastline were lost. Being the most isolated of the islands, St Agnes is inaccessible from the other islands during rough weather conditions.

St Martin's was probably the island least impacted by the storms in the winter of 2013/2014, due to its location and the wind directions (SP16: 42). Nevertheless, the storms caused more than usual erosion along the southern dunes (see Appendix A.4), where a small slipway was washed away and the old quay (which is hardly in use) was damaged (SL2: 1). Generally, the most vulnerable areas are agricultural areas behind the beaches (Fig. 5.15) and areas behind the dunes, which partly erode and partly experience flooding. In this area, the island campsite would be affected if dunes are breached (SC2: 14) and a freshwater pool could also be contaminated (field notes, 05/03/2014).

Fig. 5.15 Erosion on St Martin's, February 2014

5.1.1.3 People's Perception of Vulnerability to Climate Change

In addition to the assessment of risks described above, I consider it important to include the perception of risk and vulnerability regarding storm and flood events among the local population, to gain a meaningful understanding of the local conditions and potential implications for adaptation. On this topic, my findings refer again to the Isles of Scilly in general, rather than to specific islands.

People are accustomed to isolation in stormy weather conditions. '[T]here is also realism amongst the traditional population that if you live on a few granite islands out in the Atlantic, you can expect extreme weather' (SP3: 66). Living on an island has shaped the attitude that 'there is no getting away from the weather here' (SP10: 46) (Fig. 5.16). People who have not lived long on the islands appear to feel more affected and concerned by such problems (SP16: 70; SP9: 125; field notes, 28/02/2014).

Farmers are very aware of changing weather conditions because they rely on the weather for their economic activities. Especially flowers cultivated on Scilly, a winter crop, are sensitive to stormy weather (SC6: 37). Furthermore, the boatmen are well informed and reflective about weather conditions, as their work depends on them (SL4: 7). Generally, the impression among local people (in contrast to those in Cornwall) of the extreme weather conditions of February/March 2014 reveal that storms and flooding are surprisingly far less of an issue, and also that people are less concerned, as they are more accustomed to stormy weather (field notes, 10/02/2014; SP12: 22); as one shop owner said, 'if the flood comes, we step a little bit back, that's it' (field notes, 06/03/2014). Nonetheless, the storms were described as 'unprecedented' (SP14: 23), 'unusual' (SP13: 14), 'exceptional' (SL4: 6), 'horrendous' (SM2: 34) and 'February the fourteenth here, this year [2014], was the worst I have ever known in Scilly' (SP15: 104; cf. SP17: 27) and these events 'may have alerted more people' (SP4: 8).

The results from the household survey show that people are generally 'very aware' (Mumford 2010: 39) of the risk of sea-level rise and coastal flooding to the Isles of Scilly (Fig. 5.17), as well as climate change in general (SM1: 45; SC2: 39),

Fig. 5.16 Cartoon in a Scillonian magazine (*Source* Anon 1987: 2)

and state that they are ‘extremely vulnerable’ (SP3: 2). However, the storms that occurred just a week before the household survey began might also have influenced perceptions of the development of frequency and intensity of storm conditions over recent decades (Fig. 5.18). In contrast to the relatively clear results of the survey, with 60% of respondents saying that it has increased, and only 18% saying that it has not changed, local interview partners were more critical with regard to whether it changed or not, because ‘patterns are very tricky in Scilly’ (SL2: 7; cf. SL4: 6).

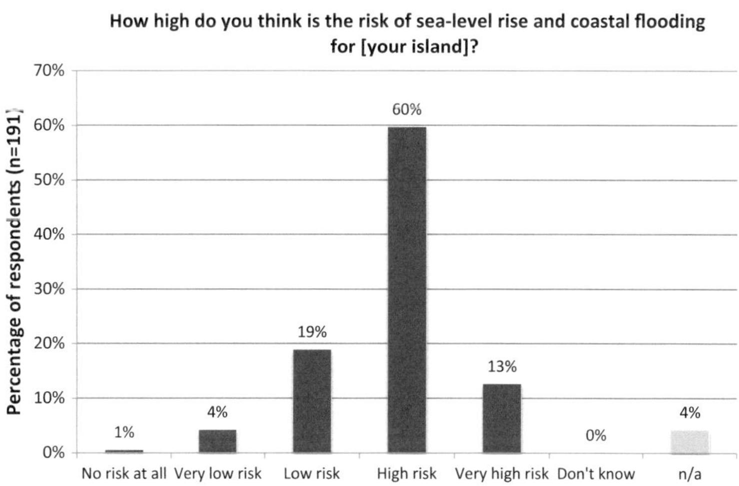

Fig. 5.17 Perceived risk of sea-level rise and coastal flooding

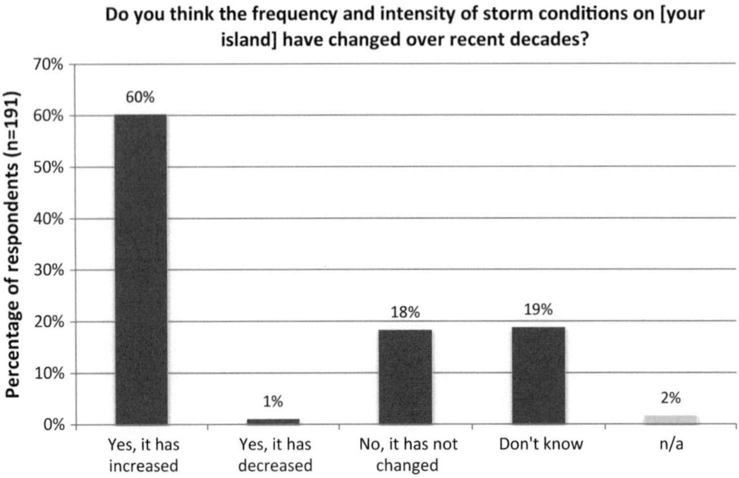

Fig. 5.18 Perceived development of frequency and intensity of storm conditions

Although people are aware of the risks and their experience of constant 'exposure to the challenge' (SP6: 34), it does not have an impact on long-term planning and 'the whole issue of adaptation to changed climate is not really deep in here yet. [The local population] tends to be reactionary' (SP6: 32). Extreme events like the recent storms cause people who are sceptical of climate change to become more concerned about it (SC2: 39; SC4: 52). Therefore, 'another disaster' is probably 'the only way that you would get anybody to respond dramatically to climate change issues' (SP1: 74) and push the council to request more funding for coastal protection from the national government.

5.1.2 *History and Society of the Isles of Scilly*

As already mentioned above, historical events have shaped people's perception of storms and their isolated, insular location. To gain a thorough understanding of the islands' current situation in terms of their political, economic and social structures, a historical perspective is insightful and crucial. For the historical development of the Scillonian society up to the nineteenth century, I draw mainly on the work of Matthews (1960), which provides the most elaborate and reliant historical account. In addition, regarding the more recent development, I refer to media and archive analysis, grey literature, as well as expert interviews and informal oral communication (recorded in my field notes).

5.1.2.1 Historical Development

Archaeological findings suggest that the Isles of Scilly 'were inhabited at least as far back as 4000 years ago' (Bowley 1980: 16). In prehistoric times, Phoenician traders used the islands as trading points and a place of call, activities that faded with the rise of the Roman Empire and a shift in trade towards the Mediterranean. In the fifteenth century, Scilly regained significance in trade, though it was especially important as a militarily strategic location against the French and Spanish and the islands underwent various phases of fortification. A key figure at that time was Scilly's first governor, Sir Francis Godolphin, whose family would hold the lease to the islands until the nineteenth century (Matthews 1960: 9–14). After peace treaties had been signed with France and Spain, 'Scilly slipped back into semi-obscurity and non-importance' (Matthews 1960: 12). This changed again during the English Civil War (1642–1651) when the islands traded hands repeatedly until the parliamentarians won the command over Scilly in 1651, followed by further military enforcement and strengthening against the threat of attacks by the Dutch.

The end of the Second Anglo-Dutch War in 1667 led to the 'birth of a new Scilly' (Matthews 1960: 15), unoccupied by military forces, which 'had left a profound mark on the character of the islanders' (Matthews 1960: 15). Matthews describes the islands as a classless community, based on kelp-making, subsistence economy of fishing and agriculture, and experiencing little poverty (cf. Ashbee 1974: 281). This period also witnessed an increase in trade, but also a rise in smuggling, due to the distance from the royal court and 'lack of proprietorial interest in the Islands' (Matthews 1960: 15). At that time, the islands were administered by the governors and proprietors of the Godolphin family, together with the so-called Council of Twelve, an assembly of inhabitants. However, the Council steadily lost influence over island policies because of the lack of clarity with regard to the extent of its power and conflicts with military commandants (Matthews 1960: 37).

The islands' socio-economic condition continued to fluctuate throughout the eighteenth century as smuggling intensified in the absence of a strong government, leading do eventual social disintegration (Matthews 1960: 89). During the Napoleonic Wars, income and employment were generated through military services once again and smuggling was definitively eradicated. Nonetheless, further social disintegration split the island communities, especially between the upper class on St Mary's and the poor subsistence farmers and fishermen on the off-islands, but also between members of the Council. Consequences of this development were the loss of self-sufficiency and the final break-up of the conciliar government (Matthews 1960: 123). However, the most important period deciding the fate of the islands was the nineteenth century, a period of 'distress and progress' (Matthews 1960: 124).

The crisis resulted from the lack of trade and loss of military importance, exacerbated by a corn and potato famine in 1817. In 1834, Augustus Smith took over the lease to the Isles of Scilly. The son of a wealthy business family, he 'possessed resources far beyond those which the islands could yield' (Matthews

1960: 166). A local historian calls him 'a social engineer', who 'tries to take the islands and make it possible for the islands to turn themselves around—to make their lives a lot more successful' (SP9: 38). 'Scilly was an ideal experimental ground' (Matthews 1960: 165) for Augustus Smith's project 'to restore prosperity and to create an ordered society' (Matthews 1960: 166). He did so by introducing several reforms, in fact, under a strict autocratic rule and probably unpopular at times, but successful (SP9: 38). After the long, negligent rule of the Godolphin family, Augustus Smith, a landlord who took a personal interest in the islands, filled the power gap in local politics. The speed and the effectiveness of his reforms are notable.

First, he introduced a local justice system, which had been lacking before. This would become the basis for all of his subsequent reforms. Scillonians were becoming 'good citizens' (Matthews 1960: 172) as well as literate citizens. The reforms of the local education system included compulsory school, longer than on the mainland, and the opening of the public library in 1870. After finishing school, one child per family would stay on the islands to take up a local craft while the others would be sent away for military service or further education (particularly the girls) (SP9: 38). The school reforms were based on the existing schools that had been established by the Society for the Propagation of Christian Knowledge (SPCK). These reforms resulted in a very low rate of illiteracy and a comparatively well-educated society (a characteristic that can still be observed on the islands today) (cf. SP9: 51).

The economy was also transformed under Smith from a poor subsistence economy, struggling with self-reliance, into a monetary economy. Smith encouraged the specialisation of businesses, such as shipbuilding, fishing, repairing and trading. Prior to the reforms, individuals had worked in multiple sectors, mainly agriculture, fishing and trading, simultaneously. In the reformed economy, shipbuilding in particular created new employment and income opportunities, bringing a substantial degree of wealth to the islands (Matthews 1960: 182). Diversifying professions also lead to a rise in inter-island trade (Matthews 1960: 199).

The flourishing economy was administered centrally by Augustus Smith. This was not universally popular, as it implied a regulated building lease system and tenure law, and it restricted the areas in which people were allowed to live and build. In this context, important infrastructure was also developed, such as the new church in Hugh Town and the quay. The islands became a thriving community, highly connected with the outside world. Plans were made for a regular postal ship from the mainland (Penzance, Cornwall) (Matthews 1960: 188). Religion was still the most important feature of social life, but more pubs and inns started to open. Due to the rapid transformation of the islands 'during these mid-nineteenth century days, St Mary's presented a scene of activity that would have been thought a miracle less than fifty years before' (Matthews 1960: 184).

In the late nineteenth century, the Isles of Scilly once again faced new challenges (Matthews 1960: 213). Firstly, the rise of steam ships greatly diminished the demand for sailing ships, which badly affected the local shipbuilding industry, causing economic decline and unemployment on the islands. Furthermore, the agricultural sector had to compete with exports from the New World. The Scillonians nevertheless found an answer to these challenges by positioning their harbour as an entrance to Cornwall and shifting their economic focus from traditional agriculture towards market-garden crops and flowers. The latter would become the basis of the economy for the coming decades. Due to the mild climate on the islands, flowers can grow earlier here than anywhere else in the United Kingdom, giving local farmers a considerable niche market advantage.

This period also witnessed a stabilisation of the population, the establishment of the first freely elected council with sub-committees for special problems (e.g., lighting, health) and the formation of a one-man police force and a private telegraph company by Scillonians. Religion remained an important social aspect, but with the introduction of cricket and other sports by Smith's descendants, the islands gained a new 'institution for the social benefit' (Matthews 1960: 228).

In 1922, the lease to the majority of the islands was returned to the Duchy of Cornwall by the Dorrien-Smith family. The next milestone in Scillonian history occurred in the 1950s, when the Duchy sold most of Hugh Town to the Council of the Isles of Scilly, also selling freeholds to islanders at very low prices. Settlements such as Macfarland's Down, near Telegraph Hill, were established. This had major implications for the social and economic structure of the islands. In time, islanders recognised the real value of their properties, sold them, and many moved away. To the Duchy, this was seen as a negative development (field notes, 12/03/2014). By this time, the economy had become very reliant on flower farming. However, it became difficult to maintain business due to rising competition from overseas. Many local flower farms ceased to exist and islanders relocated to the mainland (SP12: 96).

More recently, tourism has taken the place of the flower industry as the islands' main economic activity. As many of the properties were converted into holiday rentals or pensions 'the 1960s thought to the 1980s or 1990s were economically quite good times—people came here, they had a good standard of living' (SC3: 71). The social structure changed with the influx of people from the mainland following the departure of islanders. This new generation was referred to as the 'helicopter generation' (field notes, 23/07/2014) by one of my interview partners. The helicopter generation includes the people who arrived during the 1960s and 1970s from the mainland and bought property on the islands, which had previously been impossible when the islands were all still owned by the Duchy and the local population was more closed and family-based. Those newcomers came and flourished within the tourism-driven economy. Yet they were also accustomed to more public services and private amenities than the traditional population, which was acclimatised to the limitations of island life. A typical characteristic of members of this new generation is the reliance on and taking for granted of stable and regular connections to the mainland by helicopter (which they were also able to afford, in contrast to older farming/fishing generations). The helicopter connection was recently discontinued, upsetting many people and giving rise to the nickname (field notes, 23/07/2014).

5.1.2.2 Economy Today

As described above, historically 'people survived here by taking from the land and taking from the sea' (SP1: 30). Today, the islands' economy is so heavily dependent on tourism (cf. Grydehøj and Hayward 2014) that it is 'almost a mono-economy' (SP6: 50). Businesses from traditional occupations, such as flower farming, agriculture, fishing and boat building, still exist, but they are of minor significance in terms of employment and economic output (Ash Futures 2014b: 8). Nonetheless, the farming sector, 'underpins everything' (SC6: 25), because it is still very important in terms of shaping the cultural landscape of the islands. Farming makes 'a most important contribution to the economic and social sector [as] a spin-off for services, shops, transport, and employment' (Read 1988).

The Isles of Scilly have by far the lowest unemployment rate in England and Wales, at 1.5% (Office for National Statistics 2013). However, due to the small number of qualified specialists, recruitment of staff with specific skills is difficult if local candidates on the islands are already employed (Hargreaves 2013c). At the same time, it is difficult for non-islanders to find employment in Scilly, apart from opportunities at Tresco Estate. In fact, the estate prefers to hire locals, because people from the mainland could experience difficulties in adjusting to life on such a small island (SL1: 4). The islands also have the highest percentage of self-employment in England and Wales, at 33%, followed by the Orkney Islands (Office for National Statistics 2014). Many people have multiple jobs, due to seasonality and high costs of living on the islands. Despite the fairly upper-class market the Isles of Scilly attract as a tourist destination, 'the majority of [local] people struggle along on very little, and there is [*sic*] plenty of people who cannot afford to leave the island' (SC1: 39). Thus, paradoxically, the islands are among the poorest regions in the United Kingdom (Dugan 2008).

5.1.2.3 Stakeholders and Political Structure Today

Politically, the Isles of Scilly are a special case because they are the smallest unitary authority in the United Kingdom. The Council of the Isles of Scilly is the largest employer on the islands, making administration and support services the largest business sector in terms of employment (Ash Futures 2014b: 8). The Council is 'the most important decision-making body' (SM1: 9) on the islands. Despite 'traditionally [having] stood very much on their own two feet' (SP1: 4), today 'resources are fairly small' (SP14: 16) on the Isles of Scilly, resulting in the islands' very heavy reliance on external financial support from the national government and the European Union (Ash Futures 2014b: 9). As mentioned above, historically all of the islands were owned by the Duchy of Cornwall, who sold the land of the main settlement, Hugh Town, to the state in 1949. The rest of the land is still owned by the Duchy of Cornwall and leased out to private tenants, businesses and the Isles of Scilly Wildlife Trust. The Duchy, therefore, is a powerful but fairly passive stakeholder (cf. SM1: 23) and 'tends not to interfere, as long as they get the rent' (SP12: 86).

The local political system of the Isles of Scilly can be described as:

> peculiarly apolitical, because it is not party-based. Everybody is an independent. So, […] you haven't got that polarisation with people representing one party or another having a polarised opinion on everything. We have got many, various shades of grey. So, that can lead to never getting a really bold decision in either direction. (SC1: 24).

Consequently, people who promote an explicitly 'green' agenda by standing as candidates of the Green Party, hardly get elected (SM1: 71).

5.1.2.4 Demographics

The Isles of Scilly have always experienced fluctuations in population (SP3: 70) in terms of incomers and islanders. The total population today is around 2200 on all the islands, the land area of which is around 16 km^2 (population density: 137 people per km^2) (Table 5.1).

The demographic structure of the islands is influenced by children and young people moving away for (higher) education, usually at the age of sixteen, and experiencing difficulties finding employment on the islands afterwards. In addition, the recent economic shift towards tourism, as mentioned above, has led to an increase in 'people who have looked at Scilly as being a nice place to retire' (SP8: 106). Not all of them stay. Some people who might not have 'realised how difficult it is to get off the islands, and how expensive it is, especially in winter, to keep in touch with their families, […] they get a bit disillusioned' (SP8: 106; cf. SM3: 28; SP16: 16). Most incomers move to St Mary's, as housing is even scarcer on the off-islands, but, as one interview subject stated, they 'do not understand our [off-islands'] way of life' (SL6: 64). Often they move back to the mainland, which means that the population is very unstable. On the other hand, long-time Scillonians have also moved to St Mary's and to the mainland to benefit from better services or to find work (SL8: 60).

Furthermore, there are many second homes, kept by the owners as holiday homes and/or rented out to tourists in the summer months. However, 'it is a big problem' (SP9: 65) for the local community, because many houses are empty for most of the year, unavailable to local and working staff, resulting in higher housing prices. One interviewee indicated that this situation leads to 'a bit of resentment, too' (SP9: 67) (cf. Wakefield 1986). Therefore, there are many young people and very many old people, but few people who are economically productive (cf. SP11:

Table 5.1 Population of the Isles of Scilly (Office for National Statistics 2011b)

	Isles of Scilly (total)	St Mary's	Tresco	St Martin's	St Agnes	Bryher
Population	*2203*	1723	175	136	73	92
Area (km^2)	16.03	6.29	2.97	2.37	1.48	1.32

16). There is a perceived risk of the islands becoming 'a retirement home of the UK' (SP11: 48). In that case, many of the services could no longer be provided, including the voluntary fire services, the schools and the coast guards (SP16: 74), and 'the fabric of the islands could be at stake' (Dugan 2008). Besides incomers who move to the islands, there are many seasonal workers, especially workers in the tourism and flower industries. The former come to Scilly mainly in the summer season and many of them return each year, whereas the latter include some permanent residents because there is work year-round on the flower farms (SP16: 62). Seasonal workers and workers for the flower industry usually live in staff accommodations provided by their employer.

The overall demographic structure of the Isles of Scilly reveals an ageing population. However, the structure differs among the five islands (Fig. 5.19). While St Mary's and Bryher have a high ratio of residents over the age of 50, St Agnes, St Martin's and Tresco have a comparatively young population. This is especially the case on Tresco, although the demographics are influenced by staff in the holiday business of Tresco Estate. The other two off-islands, on the other hand, have

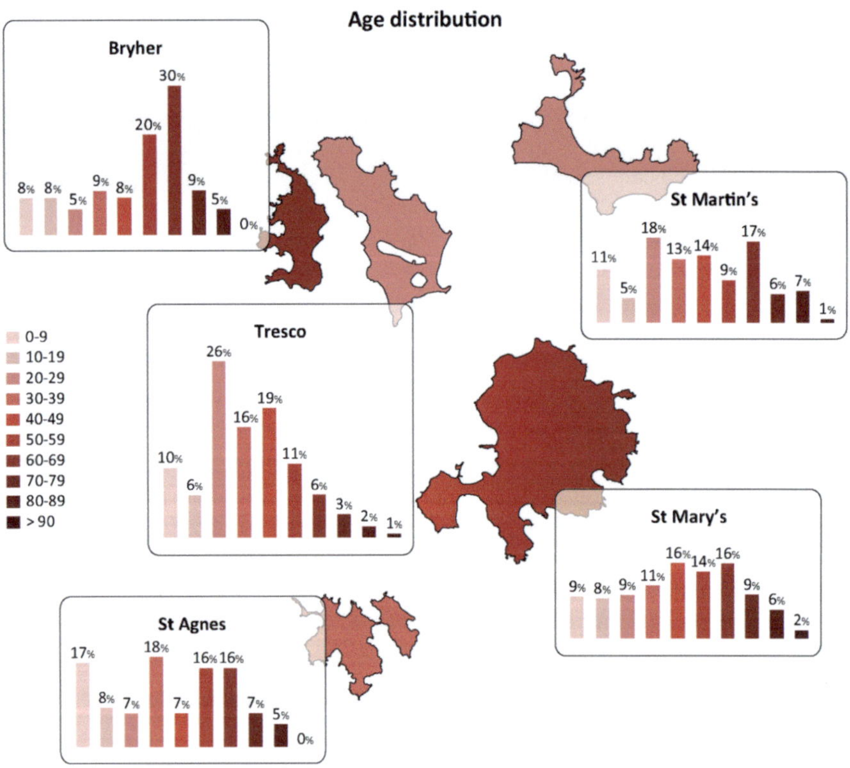

Fig. 5.19 Age distribution among the Isles of Scilly (based on Office for National Statistics 2011a)

witnessed a recent increase in newborns and young families, a situation that has been very different and problematic in the past.

There is a 'quite proud' (SP11: 62) Scillonian identity, which is based on the Cornish culture, but explicitly different (SC1: 36), for the simple fact of the islands' distance to Cornwall: 'it has always been slightly separate from the mainland' (SM1: 11). From an outside perspective, Cornwall is already perceived as peripheral. In that sense, Scilly could be seen as 'periphery of the periphery' (field notes, 05/11/2013). However, only about three per cent of the population have a genealogy of multiple generations on Scilly (SP11: 62). Furthermore, as an archipelago, 'there is a strong Isles of Scilly community feeling. And then within that, each of the off-islands has got a very singular island identity for each of the islands' (SC1: 22).

5.1.2.5 General Issues, Problems and the Future

According to the respondents of the household survey, the most important issues for the future of the Isles of Scilly are tourism and transport, followed by waste disposal, environmental pollution and food security. Climate change is, on average, the fifth most important issue (Fig. 5.20). Accordingly, also for the council, tourism has to be maintained, while also trying to diversify the economy (SP12: 12) and keeping the 'natural environment' at 'the heart of our decision-making' (field notes, 11/12/2013). The main landowner, the Duchy of Cornwall, has a clear 'hierarchy of its priorities, and on Scilly the quality of the environment is absolutely sitting at the top' (SP6: 38), not only from a conservation point-of-view, but also because the environment is the basis for the tourism industry on the islands.

Despite their reliance on tourism and the minor role of traditional activities such as agriculture and fishing, the islands are not merely holiday sites, but rather 'working islands' (except Tresco), a feature that local people are eager to emphasise (field notes, 17/02/2014). Nonetheless, the relevance of tourism to the economy is reflected by the survey results. Closely connected to tourism is the issue of transport because reliable and affordable transport links to the mainland are essential for tourists travelling to the islands, but equally important for locals travelling from the islands, e.g., for specialised medical care or for supplies of food (SP10: 50). This is reflected also in the comments from respondents in the household survey demanding improvements in the connection to the mainland (S85; S92; S101; S125). Waste is perceived as a major issue and a nuisance to many residents, primarily due to the lack of capacity of the local incinerator and the resulting accumulation of waste at the local Moorwell Waste Site, sometimes referred to as the 'Moorwell Alp' due to the large pile-up of waste there (Hargreaves 2013b). In fact, the Isles of Scilly have a significantly higher production of waste than the mainland and a lower ratio of recycling (Ash Futures 2014a: 14). Food security is an issue for two reasons. Connected to the transport issue, the supply of fresh food from the mainland via the cargo ship is not always guaranteed when there are difficult weather conditions. Yet the main problem is that the agricultural and food

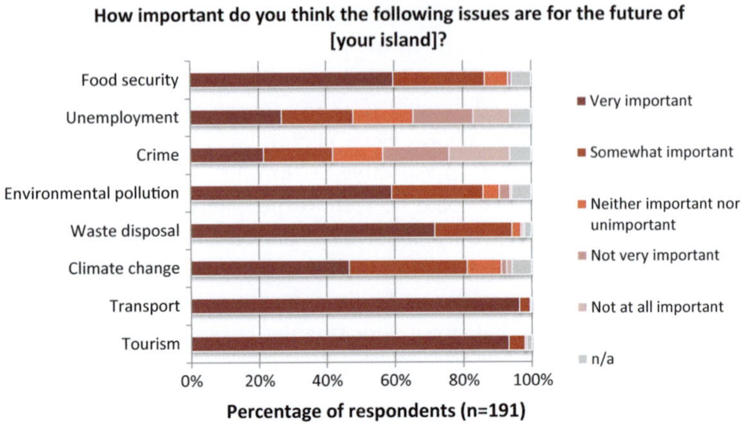

Fig. 5.20 Importance of issues

production on the islands are too small to support the community, especially during the tourism season. Thus, there is a strong seasonal imbalance: a lack of availability of fresh food for the local population in winter months, due to the transport issue and the low amount of local production, while in summer months, the increasing demand due to tourism puts further pressure on the small-scale local market (SP11: 14). However, there is potential to increase local food production (SP8: 20).

One issue that was excluded from the survey, yet proved to be critical, is housing (SP12: 14). Housing is limited on all of the islands; locals and 'imported' staff often cannot find (affordable) homes for their families. The new local plan suggests building up to 120 more houses on St Mary's over the next ten years to meet the needs of the local community and to grow and attract more businesses in diverse sectors (Ash Futures 2014b: 16).

5.1.2.6 Islands in Detail

As in Sect. 5.1.1, regarding physical vulnerability and the impact of storms, a more detailed comparison of the individual islands can also offer insights into social issues and economic and social structures.

St Mary's, by far the largest of the Isles of Scilly, both in terms of area (6.29 km^2) and population (1723), is the centre for local services and the transport hub. The main harbour and only airport, providing transport links to the mainland, are located in St Mary's main settlement, Hugh Town. Hugh Town is also the base of the Council of the Isles of Scilly and other relevant organisations such as the local offices of the Duchy of Cornwall and the Isles of Scilly Wildlife Trust, which manages most of the uninhabited land on behalf of the Duchy. The second-largest settlement is the village Old Town, followed by other settlements scattered across the island (see Appendix B.1). There is a substantial age gap on St Mary's, and there are many second homes that are empty outside of the touristic/summer season, an issue contributing to the housing problem for the local population (SP9: 65).

The off-islands are, on the one hand, considered as peripheral to the 'mother ship' (SP11: 40), St Mary's (at least from St Mary's point of view). As one interviewee explains, 'St Mary's is one step away from the mainland, and the off-islands are another step away from the island' (SP16: 86). This mentality even extends to 'off-island people […] getting the impression now that Scilly begins and ends at St Mary's quay head' (SL6: 64) for the local decision-makers, which is why, for some, 'the difference between St Mary's and the off-islands has never ever been greater than what it is now' (SL6: 66). Despite being 'each very different' (SP8: 120) and 'very separate' (SL8: 96), the off-islands are all significantly smaller than St Mary's and share challenges that are 'magnified' (SC6: 49), such as increased freight costs, isolation from the main island and one other during heavy weather conditions, and far fewer services and facilities (SP13: 18; SL8: 60). On the other hand, locals also describe them as 'very robust' (SP8: 120), 'more independent' (SP16: 74) and 'very resourceful and resilient' (SC4: 22).

Tresco has two main settlements: New Grimsby on the western coast and Old Grimsby on the eastern coast (see Appendix B.2). Tresco is different from all the islands in terms of governance because it is still completely leased to the Dorrien-Smith family by the Duchy of Cornwall since 1834 (SP18: 19). The family operate a tourism resort on Tresco and are responsible for most of the services provided on the island, including waste management, water supply and coastal protection. The single lease to the estate management also prevents people from setting up their own businesses on Tresco, in contrast to the other islands (SL8: 69). Tresco Estate employs almost all of island's residents and there is a perceived 'omnipresence of service staff' (field notes, 21/02/2014) on the island. There are, however, not many traditional local families left on Tresco. Since the island turned into one large and very wealthy estate, individual local businesses have slowly disappeared (SL8: 20). Many seasonal workers come to Tresco, often returning year after year. Another important fact influencing community structure is that it is no longer possible to retire on Tresco: at the end of their working lives, residents must leave their home, and the island, to make room for new employees of the estate (SL8: 72).

The second-largest of the off-islands is St Martin's, which is home to the largest flower farm on the off-islands, along with multiple agricultural and handicrafts businesses established in recent decades and several holiday facilities, including a hotel and campsite. The three settlements on St Martin's—Lower Town, Middle Town and Higher Town—are located along a road that stretches across the island from west to east (see Appendix B.3). There are few original islanders left, however, there has been a positive demographic change on this island with new families moving in. As one islander observes, 'things are a lot more healthy socially now […]. [M]ore people came in, and they were not so old-fashioned […]. They had a more liberal outlook on things, and they got on better' (SP12: 114).

The island of Bryher is located near to Tresco and also influenced by its economy. There is a hotel on Bryher which is owned by Tresco Estate, as are the local boat services. Most of the population lives in The Town, in the centre of the island, while there are further housing and holiday facilities, including the hotel, at the western coast of the island (see Appendix B.2). The island's economy is dominated by tourism and benefits from tourists based on Tresco. Agriculture and fishery have been practically abandoned on the island; they are certainly less relevant here than on the other off-islands (excluding Tresco). There are very few Scillonians left on Bryher. When farming became unsustainable, many of the local farmers moved away, sold their properties or turned them into holiday homes (SP15: 58–60). In the words of an islander, 'the people moved off, and there is always that kind of strange relationship between Tresco and Bryher. People have moved from Tresco to Bryher, and then to St Mary's' (SP9: 46).

The remotest of the inhabited islands, St Agnes, also has the smallest population (73 inhabitants). The settlement stretches from the eastern to the western part of the island, with three settlements—Higher Town, Middle Town, and Lower Town—very closely connected and a circular road in the centre of the island (see Appendix B.4). On St Agnes, still about 40% of the families have Scillonian roots, which is more than on any other of the islands. Most people are descendants of a few families, still recognisable by the surnames Hicks, Pender and Legg, which are common there. This is not the case, or at least less so, on the other islands. Moreover, unlike some of the others island communities, St Agnes is not ageing, despite being considered the 'most remote and traditional' (SP5: 5) of the islands. There is a 'nucleus of families that are born and brought up here. And inevitably some will go away and earn their livings elsewhere, but it continues the baseline of islander families' (SL5: 10). The island still depends on tourism, but businesses are more diverse and small-scale than on the other islands, as there is no hotel. A reason for that, according to one interviewee, could be:

> there is probably not so much wealth […] on St Agnes perhaps, to actually bring that sort of development forward. And I do not think it would be appreciated. Things are much more small scale. That is it. The island hall is slightly different, but that is not a conventional island hall. It is not just an amenity for people to play sports and have meetings in. (SP14: 29).

According to another interviewee, this social development occurred around the 1970s as people started to realise:

> that the main income for the island was going to come from tourism. And having got that idea into our minds that we could make the most of the tourist industry without it becoming a complete takeover of our lives, we could actually work as a community and have a community life, running along beside the tourism. (SL5: 10).

However, the situation on St Agnes was not always this way. Only in recent decades have the original islanders returned here with their families and created the demographic and economic structure found today, including a particular community spirit. Before that, in the 1950s and 1960s, 'it was very disjointed' and there was 'less cohesion' (SL5: 14). With the beginning of tourism businesses, standards of living increased and other businesses like cafés and the pub opened, attracting more people to settle on the island again, and providing spaces for the community to come together.

5.2 Social Capital on the Isles of Scilly

The assessment of social capital is based on the various research methods I used (see Sect. 4.3), but mainly the analysis of the household survey and semi-structured expert interviews. Significant differences between the islands can only partly be demonstrated or explained by survey results, which otherwise refer mainly to St Mary's or the Isles of Scilly in general. Therefore, in addition to analysing the survey results, I gained insights from expert and stakeholder interviews and participant observation. These enable me to describe the role of social capital on each of the Isles of Scilly, which, according to the conceptualisation of social capital (cf. Sect. 3.2.1), I will do by distinguishing between cognitive and structural elements of social capital.

5.2.1 Cognitive Domain

The cognitive domain of social capital refers to trust, norms of reciprocity, attitudes and behaviour. These elements can result from different kinds of social relationships and help to explain the emergence of collective action and the exchange of ideas and skills.

Many respondents on the Isles of Scilly stress the mutual trust among people of the island communities, because 'everybody feels very local, very linked in, very known in the community. You know, they feel, everybody knows everybody else' (SC1: 20; cf SL7: 49, SL8: 32). This is clearly reflected in the responses to the household survey from all the islands (Fig. 5.21). In fact, people do not lock their cars or houses here and the low crime rates make the Isles of Scilly the 'safest place to live in England' (Hargreaves 2013a).

Similarly, interviewees stress the general helpfulness of people in their communities (see Fig. 5.22). People help each other, especially when elderly people need assistance:

> it is quite a close-knit community. I think it is important that people are [...] looking after their neighbours and helping each other. And particularly where we have got quite a lot of elderly populations [...]. I think it is quite a tight-knit community. So, I think we have got that community spirit here, which a lot of places probably do not have. (SP10: 32).

Also, people donate generously to charities such as the lifeboat and cancer research (SC4: 10).

Furthermore, the role of families is 'still quite strong here. The social bonds are quite strong' (SC3: 53). On the other hand, in such a small community, one

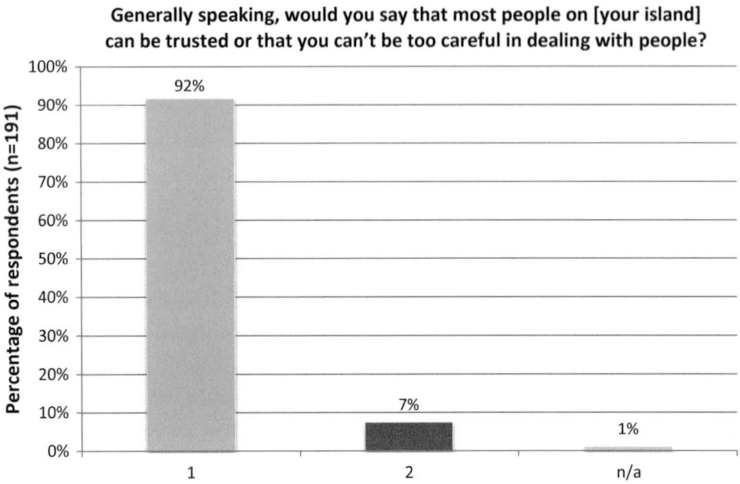

Fig. 5.21 Levels of trust

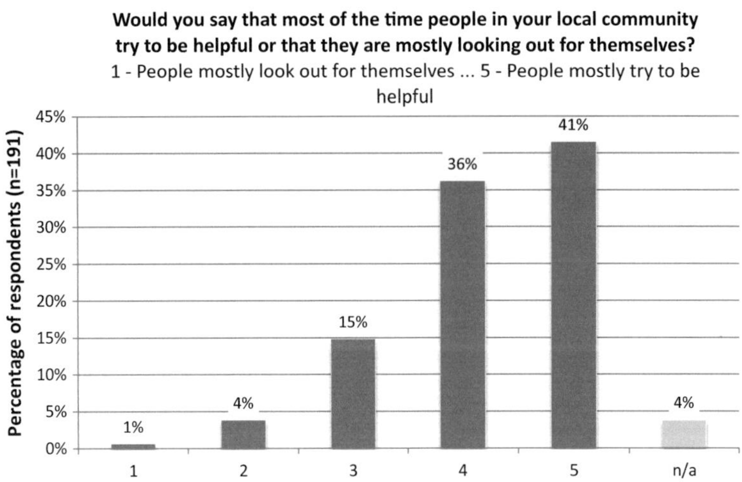

Fig. 5.22 Perceived helpfulness

interviewee explains, 'you do not want to rock the boat unnecessarily [...] [S]ome have strong opinions and express them as best they can. But you do not go out of your way necessarily to wind people up, to aggravate them.' It becomes critical when this 'social thing' leads to 'peer pressure', which 'will stop people from talking out, perhaps, because you [...] cannot afford to upset people too much, with strong opinions, because you may well need them if a problem happens. And then you do turn to your neighbour' (SC3: 53). Consequently, people are 'independent, but [...] also inter-dependent' (SP16: 58). While in general, bonding social capital and levels of trust are high among all the islands, community life, behavioural patterns and attitudes between the different islands differ and have also changed in recent decades.

On St Mary's, one can hardly speak of a single island community: 'it is not as self-sufficient, it is not as cohesive a community' (SC5: 77) in contrast to the other islands. It is a 'question of scale [...] St Mary's does not have a sense of community like the off-islands. Partly because we are too big to be one community, and too small to be lots of different communities' (SP3: 112). Another interviewee from an off-island describes the difference to St Mary's like this:

> People on St Mary's [...] tend to regard St Agnes as being what they were knowing was St Mary's like 30, 40 years ago, with that community spirit that perhaps is not quite as strong as it might have been then, on St Mary's. [...] life is a bit more hectic on St Mary's than it is here. (SP14: 25).

On the off-islands, the isolation forces people to stick together more, for example, because '[...] the boat cannot always run in bad winters. So, we do have to be a bit careful, and to look after ourselves, and look after each other' (SP14: 25). Hence, 'particularly on the off-islands, [...] where the council does not have a strong presence [...] people are a lot more independent. And there is probably a stronger cooperation because people have to work together' (SP10: 36).

One interviewee from St Agnes cogently summarises their community spirit:

> I think those who are living here at the moment all feel very happy that they are here because the community is good. That is not to say everybody gets on with everybody all the time; there are characters who do not speak to each other, those that do not fit in that well. But it generally carries along at a level where everybody that is here is happy to be here and does something towards making the community work successfully. (SL5: 12).

St Agnes's isolation pushes its people towards a certain self-reliance and to find do-it-yourself solutions (SP5: 8). This generally works on St Agnes, according to one interviewee, because there is 'a nucleus or baseline of people with a complete understanding of the island, and wanting it to stay, or continue at a sort of a business level that we know the island can support' (SL5: 6). Nonetheless, also the incomers play a role in the local community in introducing 'new ideas and new input, if you like, new enthusiasm, to continue the island way of life', as well as 'capital to invest, to bring extra business into the island [...] and to take on jobs, which other islanders, those of us that are here around, would not contemplate or would not want to have a go at' (SL5: 6).

Like St Agnes, Bryher has a small enough community to be considered a single island community, yet it shows a lower degree of networking social capital. As one interviewee puts it, 'I always say that it is a close community but not a cosy community' (SP15: 36). The characterisation of an island community can be supported by the interviewee's further statement that 'we trust each other, even if we do not like each other, sometimes, you know. It is a natural, good community. And there is all sorts of tensions between people. But at the end of the day, you know, if we had a threat from Tresco or threat from St Agnes, you know, join arms' (SP15:

38). Accordingly, also everyday work life functions in a 'competitive, rather than cooperative spirit' (SP15: 40).

St Martin's has three different settlements. One islander describes the community on the island, saying: 'if you live at this end of the island, there might be occasions, where you do not see very often certain people who live at the far end of Higher Town [… Nevertheless,] we are one community' (SP16: 22). Places like the island's pub and the island hall provide space for people to come together and the only hotel on the island provides additional jobs for younger people. In 2013, the new island hall was not yet constructed and the pub and hotel were closed, which 'had a very detrimental effect on the island' (SP16: 58), but the community came together to find ways of improving the island's situation (SP16: 84).

On Tresco, the demographic structure has changed in recent decades at least as much as on most of the other islands. A locally born inhabitant describes the island's past:

> Oh, the community was great. The community was brilliant, yes. There were a lot of islanders […] and everybody worked together […]. And we all looked out for everybody else as well. But you know, in those days – well, you still do, you still go to work together – otherwise, that's it […]. It changed quite a bit. But not for the bad, you know, but some for the good. (SL7: 29–33).

Another interviewee concurs that the community is 'really close knit […], it is a nice community to be in […]. Everybody kind of generally mixes with everybody […]' (SL8: 33).

5.2.2 *Structural Domain*

Besides the cognitive domain of social capital, structural elements are important because they facilitate sharing of information and resources and the efficiency of collective action. As mentioned in Sect. 3.2.1, some scholars have argued that the structural domain is actually the essence of social capital because it provides the structure for the development of the cognitive domain. However, a general problem of social capital theory is that it is not clear which way around it works: do people trust each other because they have the structures that bring them together? Or, do people work together because they share a common background of mutual trust? As described in the previous chapter, the cognitive domain of social capital, especially the level of trust, is very high on the Isles of Scilly. The following section focuses on structural elements, such as clubs, initiatives of collective action, civil society and integrative decision-making.

5.2.2.1 Social Networks/Clubs

To identify the basic elements within the structural domain of social capital, I listed different kinds of relationships among members of the community—friends, colleagues or other people in the community, such as church or club members—and asked the local population which of these they spend the most time with. This allows for a further understanding of the relevance of bonding, networking and linking social capital (cf. Sect. 3.2.1).

According to the household survey results, friendships (bonding social capital) are the most important relationships in quantitative terms, followed by clubs and voluntary associations (networking social capital and/or linking social capital) (Fig. 5.23). Work-related contacts are comparatively less relevant on the Isles of Scilly. A likely reason for this is that many people are self-employed and businesses are very small. Furthermore, spending time with people at church is relatively less relevant.

While historically, the church was the main place for people to come together, today this role has been taken over by sports clubs or other voluntary organisations (although church membership is still high) (Fig. 5.24). There is generally a high level of activity in clubs; the majority of respondents participate in activities in groups, clubs or organisations at least once a week (Fig. 5.25).

The sports clubs are not only competitive and hobby groups, but, as one interviewee says, they also ‘tend to be more social sports groups’ (SM1: 71). Among the sports clubs, certainly gig rowing, which 10% of the respondents explicitly claimed to be active in, is noteworthy (Fig. 5.26). Gig rowing has a long local tradition, dating back to the seventeenth century. Initially, the gigs were used as pilots for trading ships from all over the world, as Scilly was part of a major trade route from America into the Irish Sea and the Channel. Later the gigs were used as lifeboats and to rescue smuggled goods and salvage from shipwrecks, ‘a good source of income for the old Scillonian’ (Ward 1985: 10). In this way, racing between the different boats from the islands developed. In recent decades, it has evolved into a sporting event. The popularity of the sport has preserved the craft of shipbuilding on the islands, which today is a highly valued skill here. Most of the people are involved in gig racing to some degree. The rowers practice at least once a week and also race every week (at least in summer) (field notes, 16/07/2014). As a traditional activity, ‘a keen, collective thing’ (SP15: 70), gig rowing unites people of different ages, women and men, as it involves boatbuilding and maintenance, instruction, training and competition. The saying is that every child on the Isles of Scilly is born with a gig—‘gig racing is in the blood of the Scillonian’ (Ward 1985: 11). The highlight is the World Championship held annually on the Isles of Scilly. The gig rowing clubs provide not only a sporting and artisanal activity, they also create a social space that gives people the opportunity to meet others from within one’s own island, as well as from the other islands. Considering the small populations on the islands, newcomers are always welcome at the various gig rowing clubs, with their high frequency of training and races. The clubs are the easiest way for newcomers to become involved with the local community (field notes,

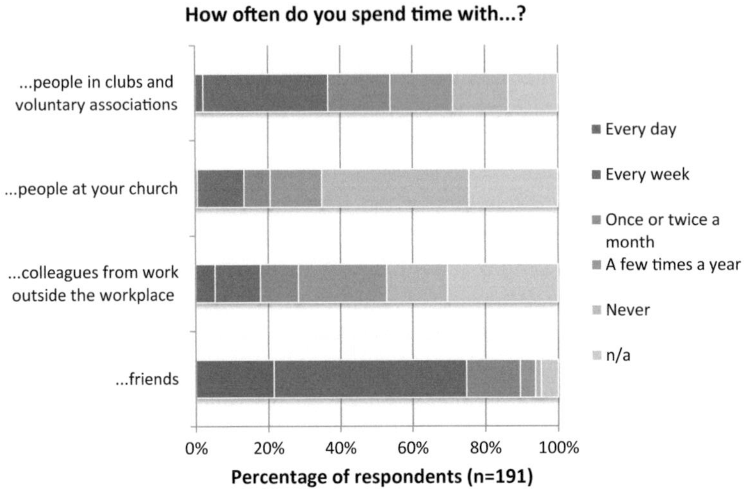

Fig. 5.23 Social activity

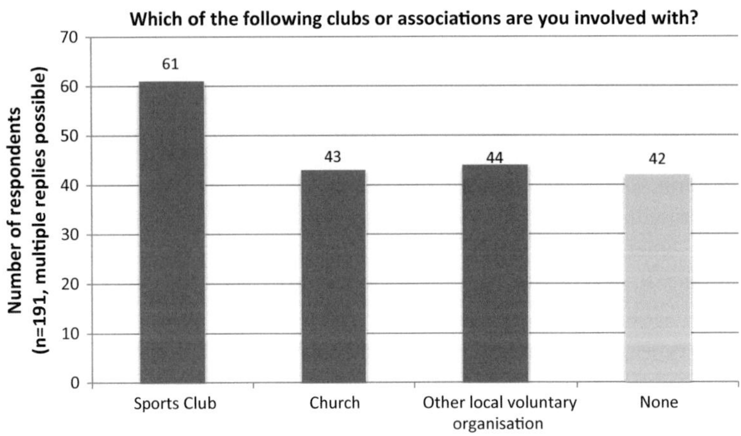

Fig. 5.24 Top three types of clubs/organisations

18/07/2014). On Bryher, for example, group activities are no longer common (field notes, 23/07/2014), with the exception of gig rowing, which is the only activity that brings people together (SP15: 78).

Another sporting activity that is popular on all the islands is cricket, which also involves more people from the community than just the actual players. In addition to forming a team, somebody has to take responsibility for the pitch, for preparing the tea (which is quite important as an occasion for getting together after the match), all the organisation of the matches and sponsoring (field notes, 27/07/2014). Similar to on the mainland UK, besides cricket, football is also popular on the Isles of

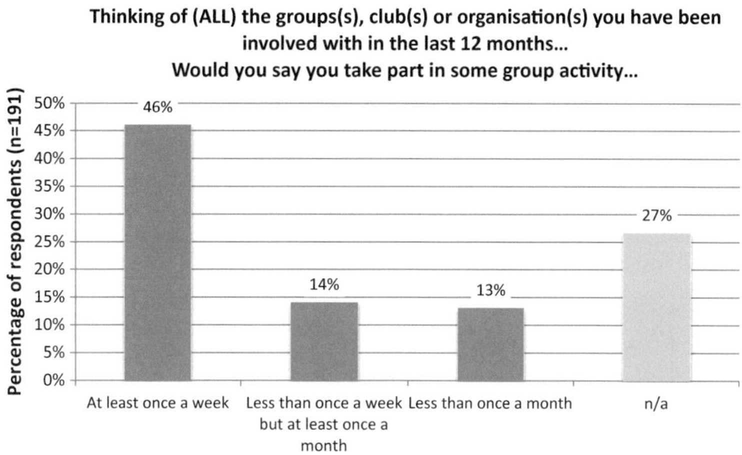

Fig. 5.25 Participation in activities of groups, clubs or organisations within the last 12 months

Fig. 5.26 Gig racing: traditional gig race (*left*) (Bryher/Tresco, July 2014); gig boat in a gig shed (*right*) (Bryher, March 2014)

Scilly, which have formed the 'world's smallest football league', with only two teams (FIFA World 2011) (Fig. 5.27). Neither cricket nor football, however, has reached the significance of gig rowing.

Other groups active on the Isles of Scilly that were mentioned in the household survey include the Rotary Club, the Women's Institute and church groups.

Fig. 5.27 Sports clubs. Cricket match on St Martin's (*left*) (July 2014); sign at football pitch on St Mary's (*right*) (September 2015)

In general, clubs and groups are the basis for networking social capital because they bring people together who otherwise might not interact. For example, on Tresco:

> you become a community here. So, as much as we are separate, […] our boys still go to the Tresco school, [and] […] sports club. […] We are all part of the community centre committee. And we do all these things. So, as a community, we still work, and see everybody […]'. (SL8: 29).

These occasions not only bring together the permanent residents, but also seasonal workers, for example, who only come in the summer months to work in the tourism businesses, often for many years in a row, and 'get involved with the community, and sports, and all that sort of stuff. And yes, you get to them and they are part of the community' (SL8: 38). One of the islanders summarises the integration of these incomers as:

> people who put themselves into the community, […] they draw, and sports events, and community events, get to know everybody in the community. So, they are sort of welcomed in and for all of us part of the community, whether they are here for six months or years. They kind of get out and put into the community. They put in and get out of it. (SL8: 44).

5.2.2.2 Collective Action

An outcome of social capital, which communities can benefit from, is collective action. Especially on the off-islands, there is an awareness among the population that many issues have to be handled by islanders themselves and that they cannot rely on the Duchy of Cornwall or the Council to solve these issues. Hence:

> community groups do step up to the plate, there are a lot of them. But also a lot of things are done informally. And if things need to be done, people club together and do it, because people are very independent […]. But when needs must, they will join together and do things. They do not necessarily expect the council to do it all for them. (SP14: 21).

Services like the fire brigade, first aid response and maintenance of community facilities are provided by volunteers from the island communities (SL5: 22).

Also on St Martin's, one islander explains, 'we are separated from the council. We do not see a lot of the council out there', and, for example, '[we] built all the roads ourselves' (SP12: 28). In fact, the roads on St Martin's and St Agnes have

Fig. 5.28 Community action: roadworks on St Martin's (*left*, *Source*: Isles of Scilly Museum [around 1960c]); construction of St Martin's Quay (*right*, *Source* Isles of Scilly Museum [around 1960b])

Fig. 5.29 Construction of Church Quay on Bryher, 1960s (*Source* both, Isles of Scilly Museum [around 1960a])

been constructed by the islanders, who are still responsible for their maintenance (SP12: 12; SP14: 27; field notes, 24/07/2014). In addition, other important infrastructure, such as quays, have not only been maintained but also constructed by islanders (Figs. 5.28 and 5.29). On St Martin's, also the habit of collective beach cleanings by the community was started, something that is now more common on almost all of the islands and contributes to the island's community spirit (field notes, 08/03/2014).

On St Agnes 'there seems to be a history of people coming together' (SP9: 46) and it 'has always been an island that has pulled together—it has always been a friendly island' (SP14: 27). The community runs a mobility service with vehicles to transport people with limited mobility, e.g., elderly people. This service is operated, financed and maintained by islanders for themselves and for visitors to the islands (SP5: 7; SP14: 65). Moreover, 'they all club together' (SC5: 79) to collectively keep the islands clean, maintain the community gardens, the quay and, most recently, also the new island hall as a major project.

Another good example of collective action is fundraising, which islanders conduct, e.g., by organising annual island fêtes to finance community projects such as sports grounds or the island halls not only on St Agnes but also on St Martin's and Bryher (SP12: 68; SP16: 96). On Tresco, most initiatives of collective action are organised by the estate management. One islander explains that, if the management says '"get that fixed", [...] everybody will jump to and get it done. [...] But, again, that is because they are being paid to do that. So, it is not them doing that off their own bag' (SL8: 89–91). Similar to Bryher, therefore, there is little cooperation in everyday life, due to the dominance of Tresco Estate (SC5: 79). Nonetheless, community life and self-organised events occur, e.g., in the island hall (which, in this case, was funded by the estate management), where people 'come together brilliantly' (SL8: 94; SL4: 4).

On a more formal level, the Council of the Isles of Scilly supports so-called local action groups (LAG) and tries to set up lobby and community groups from different sectors (e.g., agriculture, health and transport) to improve collective action and organise funding through community trusts or from external sources such as the European Union (SP11: 28). In this way, the council functions as a facilitator for rather long-term developments; yet, 'the only way to get stuff done reasonably quickly is community action' (SP11: 48), which, according to the council, can still be improved on the islands, based on the existing potential.

5.2.2.3 Participation in Civil Society

There are various NGOs and charities active on the Isles of Scilly, which enable networking and linking social capital. Classic volunteering, however, is difficult on the Isles of Scilly, because 'people are generally too busy to volunteer very much' (SC5: 49), as many rely on more than one job. Nonetheless, civil society organisations play an important role in the local council and the community as well as in environmental protection. Most of the organisations dealing with environmental protection or land management collaborate closely with each other and in cooperation with the local council. This cooperation is organised in a Joint Advisory Committee (JAC) to the local Area of Outstanding National Beauty (AONB) (SM1: 19) (Fig. 5.30). There is a considerable lack of clarity among several stakeholders, however, with regard to what AONB status implies and whether the JAC includes guiding, monitoring or even planning functions within the council (SC4: 44; field notes 11/12/2013).

The Isles of Scilly Wildlife Trust is among the most important civil society organisations—'the major player' (SC1: 14). Since 1986, it is responsible for the land management of 64% of the islands' lands on behalf of the landowner, the Duchy of Cornwall (Read 1987). The work of the Wildlife Trust includes activities, such as 'footpath clearance, clearance of bracken, installation and repairs of gates, boardwalks (…), whatever needs doing' (SC4: 4), but also environmental education and relocation of coastal paths in cases of erosion (Read 1998a). The work is carried out by employed staff members with the help of volunteers, both local and from the mainland (field notes, 05/03/2014). However, the 'set up is very, very different' on the islands, in contrast to the mainland activities of the Wildlife Trust,

Fig. 5.30 Meeting of the Joint Advisory Committee (JAC) for the AONB, Hugh Town, 11/12/2013

where volunteers are explicitly recruited and supervised. On the Isles of Scilly, says one interviewee:

> if people want to do something for their environment, or their community, they just go and do it. They do not necessarily need to be part of the Wildlife Trust or part of something. So, they just crack on and do it. (SC4: 6).

Activities such as beach cleaning that are facilitated by the Wildlife Trust with tools and material have 'worked pretty well' (SC4: 8) and involved relatively many local helpers compared to similar activities on the mainland.

In recent years, however, there have been ongoing tensions between locals and the Wildlife Trust because some residents have a 'strong opposition to their land management policy' (Hargreaves 2015c). Occasionally this opposition is expressed through vandalism, e.g., tearing down fences around animal grazing areas or damaging boardwalks set up by the Trust, which, in the local people's opinion, 'destroy' (Mumford 2009a) the landscape (cf. SP9: 83; Read 1992; Gibson 1997a: 3).

An important and successful project, which was a collaboration between the Royal Society for the Protection of Birds (RSPB), the Wildlife Trust, the Duchy of Cornwall and the community of St Agnes, was the rat eradication programme, which would not have been possible without the support of the local population

(SC5: 43, 57). The RSPB is an organisation that generally relies on the support of the community because the birds' habitat is highly sensitive. Locals, e.g., the boatman, who are constantly around the birds, have great knowledge and information about them, which the RSPB can benefit from. Additionally, the RSPB, as a national charity, can draw on external funds (SC5: 5).

Another active organisation is the Islands' Partnership, which seeks to promote business on the islands, especially tourism businesses (of which 300 are currently promoted), as this is the sector from which the Partnership initially developed. It has a 'strong partnership' with the AONB unit and the Wildlife Trust (field notes, 11/12/2013). However, it is a young organisation, which still must prove how efficiently it can promote what is effectively the majority of the islands' business interests (SP8: 68). Moreover, as tourism is the main source of income for all the islands, the Partnership is an organisation that also brings together businesses from all the islands and seeks to 'encourage people to work together' (SP8: 68). Such sectorial cooperation is rare (SL6: 56), but it has increasingly become the norm in the farming and agricultural sector, as will be explained next.

The work of civil society organisations and cooperatives such as the Isles of Scilly Farmers' and Growers' Initiative[1] (IOSFGI) exemplifies the connection of a cognitive element of social capital (trust) with the structural domain. The Farmers' Initiative's work depends heavily on gaining the farmers' trust. In contrast to similar initiatives on the UK mainland, the Farmers' Initiative operates in an environment where, as one interviewee says, 'it is different over here, because you know everybody anyway', which makes it 'a lot easier to get [the farmers] to trust us, because we know them all. And my colleague [...] is a farmer herself. So, we do not have to start going talking people around, because they are already there, pretty much' (SC6: 11). The Farmers' Initiative is funded by the Department for Environment, Food and Rural Affairs (DEFRA) and the European Union. It organises workshops, collective growing and funding schemes, exchange of information with similar initiatives on the UK mainland, or trips to other farms, e.g., to the island of Alderney to gather ideas for how to set up a communal abattoir (SC6: 19). For such projects, not only are collective and external funding sources needed, but it is also necessary 'to get people to actually physically put the thing up [...] and [...] persuade the farmers to give their time and their energy' (SC6: 33); it 'is a good way of farmers meeting up and talking' (SL8: 112). Besides facilitating collective action by the farmers, the Farmers' Initiative also tries to bring together the farmers and the rest of the community by showing that farmers do more than just providing food and flowers: they also act as 'custodians of the landscape' (SC6: 61), which, after all, is Scilly's greatest asset and the foundation of its tourism industry. Representing many businesses involved in agriculture and food production, the Farmers' Initiative cooperates with the AONB and the other conservation bodies (SC6: 4).

---

[1]According to the common use on the Isles of Scilly I will in the following refer to the IOSFGI as the *Farmers' Initiative*.

A less formal organisation is the Duchy Tenants Association (DTA), formed in 1996 to strengthen the tenants' voice in dealings with their landlords (McCarthy 2006). Besides uniting tenants, businesses and farmers from the Isles of Scilly, the association has also brought in consultants from the mainland for external expertise on issues related to tenancies.

A small environmental NGO that is active on the islands is Transition Scilly, founded in 2006, and comprising just a few islanders. The organisation seeks to raise awareness of climate change and general environmental concerns and to promote renewable energy schemes. For these purposes, the group established the Isles of Scilly Renewable Energy Cooperative (ISREC), which arranged for consultants to visit the islands, and raised funds for a solar panel on St Martin's. The latter project was abandoned, however, due to lack of cooperation by the regional power supplier (SC2: 39).

5.2.2.4 Integration in Decision-Making

Integration of the community in decision-making is officially ensured by the election of representatives in the local council—the councillors. An additional means of involving the community are the public consultation meetings and council sessions, which are usually open to the general public as well. In fact, as one islander observes, 'any [topic] what you consider to be a fairly small project on the mainland, is generally consulted with the general public' (SP1: 66). Formally, this indicates a high integration of the community in decision-making. A member of the council states that 'you get almost consultation burn-out' (SP1: 66). In fact, while almost half of the respondents to the household survey attend public council sessions or consultation meetings, most of those who do, did so only 1–3 times in 2013[2] (Fig. 5.31).

The responses to the question of why one does not attend these meetings can be grouped into six different categories: lack of time, e.g., due to work (24 respondents); inconvenient accessibility, especially from the off-islands (13); general lack of interest in politics (9); useless to attend/critique of council (8); lack of advertisement (8); and health or age constraints (7).

The attendance of council sessions and the degree of direct influence of members of the community on the decision-making process also depend on the topics and issues discussed by the council. For some 'contentious' (SC1: 28; SP3: 120) issues, i.e., 'when it directly affects them', as one islander observes (SM3: 56), spontaneous community groups can form and be quite active, e.g., regarding the Shoreline Management Plan, the redevelopment along the beach of Porthcressa or the handling of the waste problem on St Mary's (field notes, 12/12/2013). According to an interviewee, 'people have been very vociferous' about the latter topic and '[t]here

[2]Ordinary council sessions are accessible as a live stream online, but there are no statistics available about how many people follow them in this way.

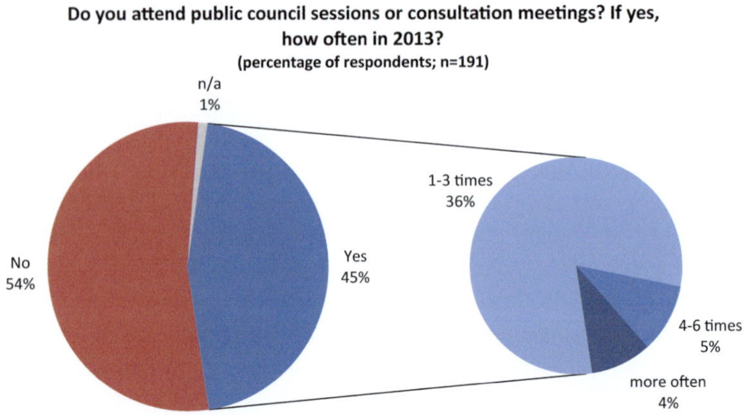

Fig. 5.31 Attendance of public council sessions or consultation meetings

have been campaign groups set up […] across the islands, different communities. Everybody has an opinion on that' (SM1: 49). To avoid losing connection with the population:

> the council every year, or every couple of years, goes out to the community and asks them what their key issues are. And even though that generally tends to tell you exactly what you already knew, we have got that consultation going on all the time to make sure what the local authority is doing is actually in line with what people want. (SP1: 70).

Moreover, the AONB is planning to set up a forum group for the local community, not only to receive feedback and give consultation on specific projects but also to maintain a permanent exchange with the local population (SP4: 26).

In the past, there has been a split between the electorate and the council officers. '[T]he perception from the community was, that there was very little point in actually engaging with the council because they would do what they liked anyway' (SP8: 70). After a change in the executive position in the council in 2013, there is now 'hope that we would be moving into more of a consensus-based politics – moving forward' (SP8: 70). The community group HEART (Honesty, Ethics, Accountability, Respect, and Truth/Transparency) was formed in 2012 to 'encourage a better standard of engagement with the community' (SP8: 74). According to external consultants as well as stakeholders from local authorities, the potential of community-based initiatives and public participation in local decision-making has not been exploited enough in the past; but it is now seen as an opportunity to increase active collaboration, exchange of knowledge and trust between stakeholders and the population (cf. Pound 2004; Isles of Scilly AONB Unit 2010; Ash Futures 2014b). Overall, however, the majority of respondents to the household survey agree that people on the Isles of Scilly can influence decisions that affect their community. This is important, e.g., to 'discuss and put forward our case for the actions that the bigger organisations, the council, the Environment Agency, have on their agenda—to […] at least make a representation' (SL5: 30) (Fig. 5.32).

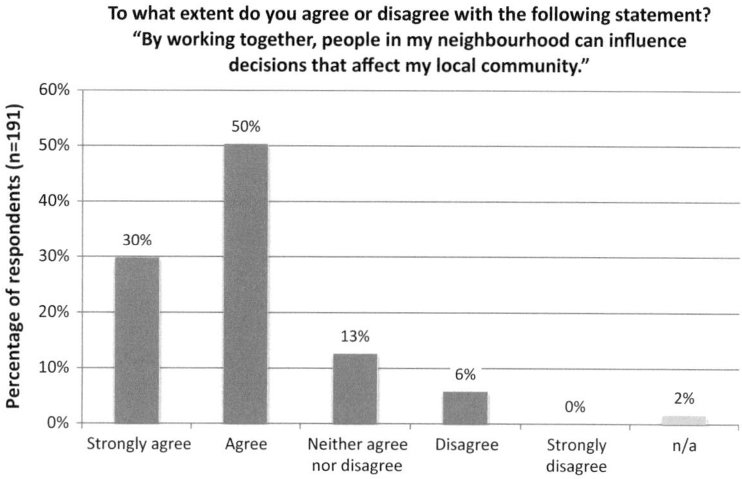

Fig. 5.32 Ability to influence decision-making

As mentioned above, the political representation of the community is provided through the elected councillors. Each off-island has two councillors representing the population in the Council of the Isles of Scilly, in addition to eleven councillors from St Mary's. However, when it comes to integration in planning and decision-making, there is a gap between the off-islands and St Mary's. Some interviewees, in fact, state, 'St Mary's is closer to Truro than Tresco' (SP2: 4), and 'they are a council of St Mary's. They do not understand the off-islands' (SL6: 40).

Tresco, as emphasised by several interviewees, is a special case, because 'Tresco is basically Tresco Estate' (SP3: 116). Decision-making works more in a top-down manner here than on the other islands, due to the central management of the estate, e.g., 'if you want any building or anything like that, the management does it, that's it. We are not allowed to say yes or no' (SL7: 71). Nonetheless, the potential of collectively influencing planning and decision-making is in some cases even higher on Tresco than on St Mary's because people can directly address Robert Dorrien-Smith, who runs the privately owned estate. If people unite and raise a certain issue, he is more likely to be active on that issue. Then, as a private business, things can be realised relatively quickly and efficiently, without much of the bureaucracy involved on the other islands (SL4: 4).

5.2.2.5 Sources and Flow of Information

People feel generally quite well informed (Fig. 5.33), not only through media or official sources from the council but also simply by the fact that people all know each other. Information is passed on very quickly and informally through word of mouth (SP1: 68).

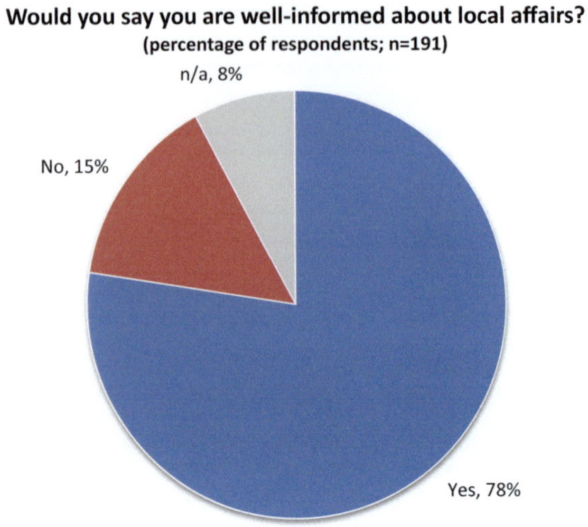

Fig. 5.33 Perceived level of information about local affairs

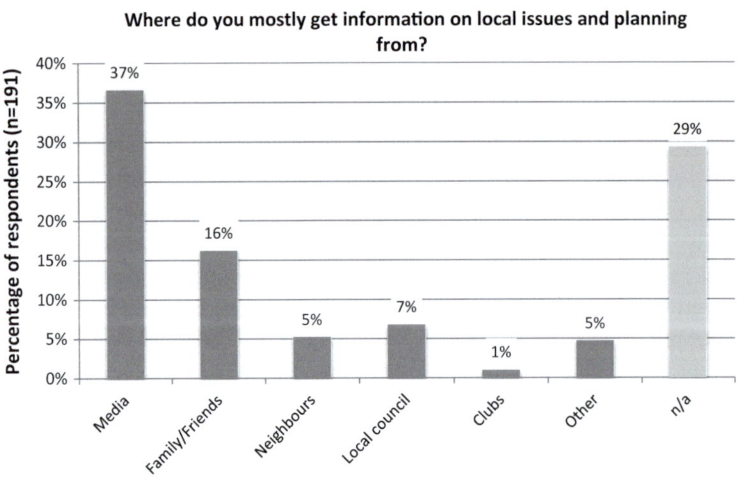

Fig. 5.34 Sources of information

According to the household survey, the main source of general information for people on the Isles of Scilly is media, followed by family and friends, and the local council (Fig. 5.34). In terms of media, the community radio station *RadioScilly*, launched in 2007, is the primary source of daily information regarding local issues, both over the radio waves and online. The station's website doubles as a discussion forum, often a critical resource for the islanders: '[p]eople are very vocal on Facebook. They do a lot of commentary [...] about all aspects of council activity, inactivity, failed to deliver services or perceptions that there is far too many people in the council doing nothing' (SC1: 30). Apart from the RadioScilly website, various other Facebook groups cover information about the Isles of Scilly and are used for discussions. Many people participating on such platforms, however, are not islanders, but rather tourists who come to the islands regularly.

The council uses its website as well as notice boards on the islands to announce new reports, upcoming consultation meetings, etc. Communication between friends, family members and neighbours, however, is not surprisingly the second-most used source of information because, as stated above, everybody knows everybody on the islands, which also leads to a high degree of gossip (SM1: 65). Classic print media exists on the Isles of Scilly only in the form of the monthly magazine *Scilly Now & Then*, the quarterly *Scillonian* magazine and a weekly column in *The Cornishman* (based in Penzance, Cornwall) by a local journalist. In summary, people rely on one major local media source, which can be problematic if the source is biassed. In fact, the council can feel misrepresented very easily and not everybody agrees with the communication from RadioScilly (SP9: 137).

5.3 Adaptation on the Isles of Scilly

As presented in the Sect. 5.1.1, the specific islands of the Isles of Scilly have different vulnerabilities towards climate change and each suffered differently from the storms in the winter of 2013/2014. The main issues regarding climate change adaptation are 'about response to emergency situations, but also [...] managing our coastal defences, making sure they are as strong as possible. And then really planning ahead for the long term, which is fifty to a hundred years' (SP1: 18–20). How do the islands' authorities deal with the challenge of climate change? How do people cope with storms, floods and rising sea levels, and what role does local community action play on the islands?

The following section analyses formal adaptation planning and strategies and informal activities by the communities on the Isles of Scilly in the context of climate change and associated pressures on the islands. There are differences between the various islands and I will, as in Sect. 5.1.1, focus on the reaction of the communities to the events of the winter of 2013/2014, which have been comprehensively recorded especially by the qualitative research methods applied.

5.3.1 Formal Adaptation and Perception

Generally, the issues of climate change and sea-level rise are including in the agenda of the Council of the Isles of Scilly. However, responsibilities and a clear strategy have only recently been formulated (Mumford 2009b). The change in national government in 2010 led to a release from the statutory requirements for local councils to explicitly invest in climate change adaptation (SP1: 4). Still, most measures implied by climate change adaptation, such as coastal protection, are 'part and parcel of what we do on a day-to-day basis [...] [and are] driven a lot by central government requirements, EU requirements etc., as well as our local response to the environment' (SP1: 74). In the case of serious damages to coastal protection infrastructure, the council relies on national authorities to provide funding for possible major repair works or improvements. Funding has to be discussed with the Department for Environment, Food and Rural Affairs (DEFRA); planning mostly involves external consultants and contractors and materials and workforce are shipped in from the mainland. Consequently, major coastal defence works have always taken very long to realise (cf. Mumford 1993, 1996). Besides funding for sea defences, the council also administers flood support schemes to help individual properties and businesses, which are affected by flood damages (Council of the Isles of Scilly 2014).

5.3.1.1 Coastal Protection and Climate Change Adaptation Planning

The basis for climate change adaptation in terms of coastal protection is currently the Shoreline Management Plan (SMP) for Cornwall and the Isles of Scilly, in its second revision (Royal Haskoning 2010). The SMP assesses the islands' vulnerability to coastal risks from sea-level rise and erosion, evaluates existing levels of protection and develops policy recommendations for the short-term (0–20 years), medium-term (20–50 years) and long-term (50–100 years).

Generally, there are four different strategies for coastal protection involved in these recommendations:

- No active intervention (NAI): No investments in maintaining coastal defences or the natural coastline;
- Hold the line (HTL): Maintain or upgrade the level of protection provided by coastal defences or the natural coastline;
- Managed realignment (MR): Manage realignment of the natural coastline, either seaward or landward towards a future sustainable shoreline position;
- Advance the line (ATL): Construct new coastal defences seaward of the existing defence line.

The SMP suggests NAI along undefended and uninhabited coasts and HTL of the existing defences as short-term strategies. In the medium and long term, however, MR is the preferred plan recommendation for many of the sea defences, especially in Hugh Town, St Mary's (Fig. 5.35), where key assets are affected (Royal Haskoning 2010: 33). ATL is no longer suggested at all in the second SMP. On St Martin's, where there are no hard sea defences, the SMP suggests NAI along the entire coast because of 'very little risk identified', while 'low rates of shoreline recession' are 'likely' (Royal Haskoning 2010: 38), and NAI satisfies AONB designations. Similar to St Mary's, on St Agnes, Tresco and Bryher, NAI

Fig. 5.35 Town Beach: around 1910 (*left*; *Source* Isles of Scilly Museum [around 1910]) and today (September 2015)

and HTL are the suggested strategies, with MR selectively on Tresco, where the main hotel is affected.

Besides explicit policy recommendations, the SMP suggests monitoring and surveillance of cliffs, dunes and beaches. However, it lacks consideration of increased peak tides and storm surges in its future projections (field notes, 11/12/2013). It also just negligibly mentions the risks to agricultural areas and that 'the policy of NAI through erosion may cause disturbance or deterioration to historic sites and their settings' (Royal Haskoning 2010: 53). Furthermore, despite the introductory statement of 'one of Scilly's greatest assets' being the 'extremely strong sense of community and [...] cohesion and sense of identity within the local population' (Royal Haskoning 2010: 1), the report neglects the role of the community for coastal protection and adaptation.

In fact, although 'certainly all through [...] the centuries islanders have repaired problems on the coast' (SC2: 27), today, coastal protection 'is really being driven by the authority, rather than the people. [...] people do not tend to be involved in that' (SM1: 43). For the protection of Hugh Town, for example, larger scale hard infrastructure is necessary, which can hardly be realised by local collective action (SM3: 54). In addition to its statutory responsibilities, the council deals with individual protection from flooding by providing consultation for local businesses and residents (Council of the Isles of Scilly 2010: 34). The Environment Agency monitors and plans the larger scale coastal defence works, but also offers the service Floodline Warnings Direct, which issues flood warnings to islanders on their mobile phones.

Also the Duchy of Cornwall and the Isles of Scilly Wildlife Trust deal with coastal issues. The Duchy is the main landowner, but tenants, often farmers, are responsible for protecting their own housing. The Duchy is conservative in investing in coastal protection on Scilly (Mumford 1993) unless key infrastructures or monuments are affected. Coastal protection is a matter of priorities, which currently lie elsewhere for the Duchy (SP6: 46). It allocates its 'resources where they are going to be of greatest value. And in most cases that will not be to protect farmland' (SP6: 24). Most recently, however, the Duchy improved the sea defences in Porthlow to guarantee better protection for the mostly residential properties in the area (Hargreaves 2015a). The Wildlife Trust does not actually deal with sea defences because the 'sea is just doing what it does, so, it will in a way be futile to try and stop it; particularly in areas where that is not where people live [*sic*]' (SC4: 14). Therefore, their work rather involves taking care of paths and dunes that are eroding.

Consequently, as with the suggestions in the SMP, structural coastal protection is usually only implemented for the protection of valuable property or critical infrastructure. Many other parts of the islands, such as smaller residential premises, agricultural and gardening areas and conservation zones, are not covered by this, and the according strategy here is NAI or MR (cf. SP2: 3). One of the most important protection measures was the 1995 construction of a concrete seawall replacing a very vulnerable wooden structure along Porthcressa (SP1: 26) (Fig. 5.36), the area where the February 2014 storm hit the town hardest. Similarly, in Old Town, seawalls have been recently reinforced and raised (Fig. 5.37), but they are still likely to be insufficient in case of similar events in the future. Hard sea defences, such as dykes, have also been built on some of the off-islands in the past, but there is basically no official ongoing maintenance, nor are any further projects planned for the future (cf. Ash Futures 2014a: 20).

In 2010, the council developed a climate change strategy for the islands (Council of the Isles of Scilly 2010).[3] It builds on the SMP but considers further impacts of climate change on the islands such as transport disruptions, loss of agricultural land and biodiversity, and freshwater shortages. The strategy suggests focusing also on natural sea defences, such as dunes, in addition to planning for necessary managed retreat. Due to the relatively low chance of national government funding for the Isles of Scilly, the council highlights the necessity of cooperation of the most important stakeholders (the AONB, Wildlife Trust and Duchy of Cornwall) for planning, funding and implementation of such projects. Important aspects to consider are also the integration of adaptation and mitigation strategies into the islands' general development plan for economy and housing, and the consultation and information of the local community (Council of the Isles of Scilly 2010: 33–34).

In contrast to coastal protection measures, which mainly address protection from storm surges, and, as mentioned, are already on the agenda, long-term climate change adaptation must involve 'transforming our settlements' (SP10: 18) because a managed retreat already seems inevitable in some vulnerable areas of the islands today (SP11: 46). This mainly implies allocating new housing and relocating key facilities to higher areas, e.g., around Telegraph Hill on St Mary's. For such long-term planning, the council collects and records data and information about the islands' vulnerability and it has already published a climate change strategy.

[3]Meanwhile, the document is not available anymore on the council's website. Moreover, the council does not employ a climate change officer anymore, after a staff restructuring in 2015. I was, however, not able to get a statement by the council regarding a possibly changed position towards climate change adaptation.

Fig. 5.36 Seawall at Porthcressa (from *left* to *right*): Wooden sea wall in the 1960s (*Source* Isles of Scilly Museum [around 1960d]), concrete seawall under construction in 1995 (*Source* Gibson 1997b: 26), redeveloped beachfront in 2014

Fig. 5.37 Raised seawall in Old Town, December 2013

Besides residential areas, facilities or infrastructure, the farmers are considered to have become 'pretty well adapted down here to storms anyway' since the nineteenth century, by cultivating small fields protected by high hedges and avoiding the areas most exposed to strong winds or erosion (Fig. 5.38; SC6: 41).

Projects that involve climate change mitigation measures include different kinds of renewable energy sources. Wind power and photovoltaic elements, which have been in use in the past and are promoted by several people on the islands, have not been possible to establish, mainly due to conservation concerns. A great hope shared by all of the islands' major organisations, including the landowner and the council, is the installation of wave power structures, which provide carbon-free energy and do not affect the natural environment or landscape, while providing employment and income opportunities for the local community (Hargreaves

Fig. 5.38 Agricultural fields protected by hedges (September 2015)

2015b). A further strategy put forward by the council to cut emissions is the promotion of electric cars (Hargreaves 2014a).

5.3.1.2 Relocation as an Adaptation Strategy

Relocation within the islands and the construction of new settlements in higher areas, as suggested by the council (Council of the Isles of Scilly 2010: 34), is difficult because of the limited space available and the high administrative barriers, e.g., due to the AONB, SSSI, EMS and Scheduled Monuments designations (SP10: 18).

The household survey shows that the majority of the respondents cannot imagine moving due to risks of sea-level rise or local flooding (Fig. 5.39). However, 39% would 'maybe' be prepared to move and 5% even 'certainly'. Thus, there is an awareness, especially on St Mary's, that 'if [the sea level] goes up much in Hugh Town, we would have to [move], unless we become amphibious' (SP9: 123).

5.3.1.3 Perception

The perception of climate change adaptation and coastal protection differs considerably among the islanders. Compared to residents of coastal areas in South West England, some respondents (e.g., SP1; SL6) see the Isles of Scilly in 'the strongest position of anybody to survive' extreme weather events because:

> we have now got quite significant sea defences and quite a lot of adaptation to protect local properties and local infrastructure, or respond at least to the emergencies that arose. Whereas on the mainland, they have carried on building, and there is now a lot more of very valuable infrastructure and properties, and industry, all right up on the coast. (SP1: 26).

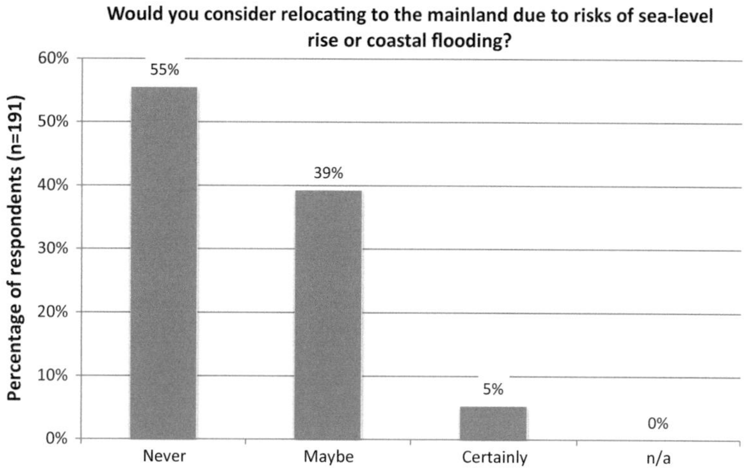

Fig. 5.39 Readiness to relocate due to risks of sea-level rise or coastal flooding

Thus, in the short term, the islands are 'quite well up on how to adapt to storms from the sea and what needs to be done' (SP1: 26) and 'historically […] we have never had as good as sea defences' (SL6: 22). In general, however, Scilly is 'a microcosm' that 'has an intimacy about it, which is quite delicate, in as much as [if] we get it wrong we will foul the whole thing' (SP6: 52). In the long term, adaptation to sea-level rise is rated as 'very poor' (SM1: 31) and there is a feeling that 'a more radical solution' (SP3: 6) must be found. A conservationist described Hugh Town as 'one large beach with a few boulders' (Mumford 1993: 76). Accordingly, overall flood and coastal protection are perceived as 'mediocre' on average by the local population (Fig. 5.40) and most people (85%) would support extra spending by the local council for flood and coastal protection.

The construction of seawalls that could function as long-term adaptation measures is seen by the islanders as a task for the local authorities and the Environment Agency (SL5: 26). The majority of people (76%) think that the local authorities should be more active in protecting the islands from the effects of sea-level rise and coastal flooding (Fig. 5.41). Individual islanders, however, are also sceptical about solutions designed by external experts 'coming down from London or elsewhere and telling Scillonians what to do' (Read 1985) without knowing the islands as well as locals (SP15: 88). In response to the household survey, several respondents (S6; S16; S20; S27) call for more 'sympathetically' (S16) designed and adapted solutions, rather than technical infrastructure, such as seawalls.

The problem with long-term adaptation is that:

> we probably will not get funded because there will be places that have higher populations and are at higher risk, economically, than Scilly, that are likely to get the money for any sort of sea defences. So, I suspect there would just be a sort of managed retreat here if anything does change significantly. That worries me. (SM1: 33).

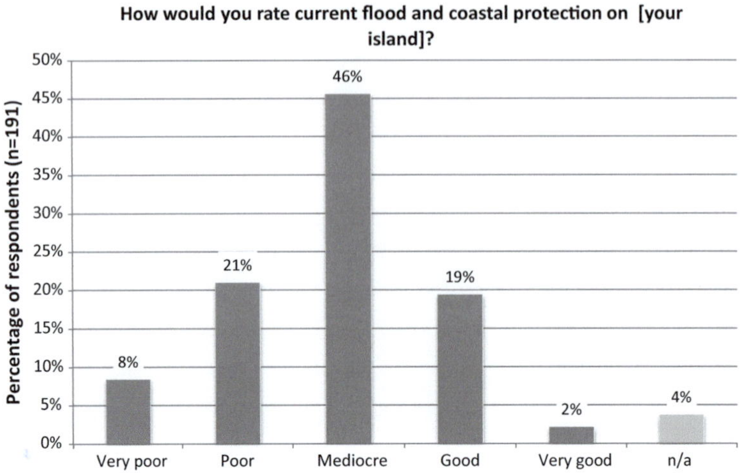

Fig. 5.40 Rating of flood and coastal protection

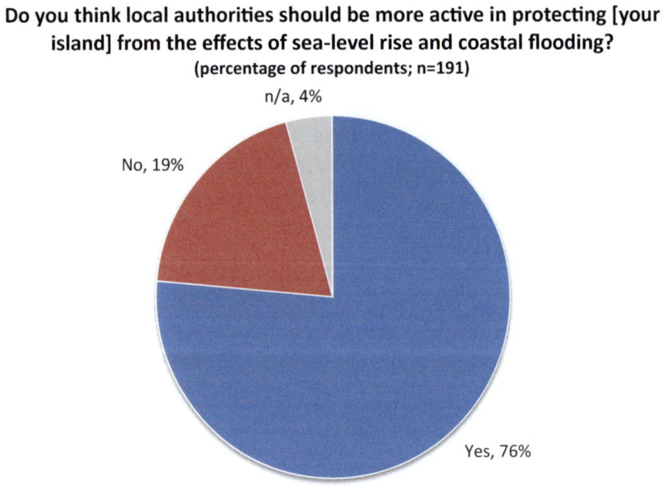

Fig. 5.41 Expectation from authorities for coastal protection

> Us, on a tiny little island out here – it is not priority. (SC4: 20).

In fact, when asked whether people expect more support from the national government to protect the Isles of Scilly from the effects of sea-level rise and coastal flooding, 30% answered ‘no’. Almost a third of the population not expecting support from national government can be considered high, given the background of recent storm events (cf. SP9: 139). However, one worried local also realises that:

> there is absolutely nothing we as a community really can do to protect our island [...]. As a little island community, we have not got the resources or finance, or plant and machinery, to really have much impact on long-term projects. (SL5: 26).

Hence, adaptation measures by the local population entail 'crisis management, rather than long-term planning' (SL5: 26).

5.3.1.4 Information

Regarding extreme weather events, people feel well informed, as they are accustomed to observing weather conditions as part of their life beside the sea. Especially in winter, daily changing weather conditions influence transport links to and between the islands. In the case of extreme weather conditions and warnings by the Environment Agency, the council officers and the councillors spread information, e.g., via the local radio, and organise protection measures (field notes, 16/02/2014). In addition, the local radio reports on local flood events and informs the public of the local council's actions.

In addition to the council and NGOs like Transition Scilly, the event E-Day, which was organised on the Isles of Scilly in 2009, has contributed to an increased awareness of the risks of climate change for small islands. Environmentalists, politicians and scientists were invited to share knowledge and information about risks and adaptation strategies (cf. Hickman 2009).

Regarding long-term adaptation to sea-level rise, however, there is still a lack of information and a sense that 'more should be done to include the community [...] in the process of decision-making. They should have the opportunity to learn a bit more and to kind of understand a little bit more as well' (SP4: 24). This is partly reflected in the responses to the household survey, where the majority of respondents claimed to be 'neither well nor poorly informed' regarding the impacts of sea-level rise. While 26% feel 'well informed' and 8% even 'very well informed', 20% feel 'poorly informed' and 7% 'very poorly informed' (Fig. 5.42).

5.3.2 Community-Based Adaptation

As described in the previous section, the suggestions of the Shoreline Management Plan and the activities of the major players on the Isles of Scilly do not have a very encouraging effect on the local communities on the Isles of Scilly. People are dissatisfied overall with the current perspective for vulnerable areas on the islands but are aware of the impacts of storms and sea-level rise. The following section focuses on activities that involve or are even initiated by members of the community, either in general or during extreme weather events like the storms during the winter of 2013/2014.

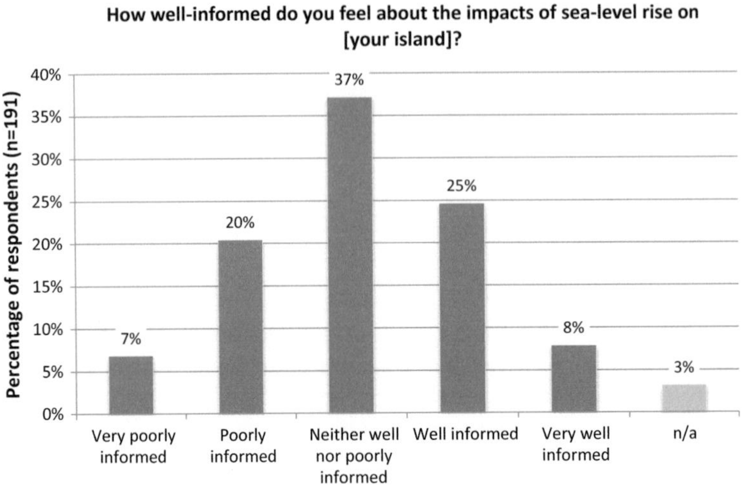

Fig. 5.42 Perceived level of information regarding the impacts of sea-level rise

The most relevant actions people are collectively involved with are emergency response and flood protection (e.g., with sand bags), but also the maintenance and repair of coastal defences, as well as beach cleanings or clearance of debris after storms. The various sources investigated during my research consistently demonstrate the relevance of collective action in this context, in particular, with regard to the events that occurred before and during my research stay on the Isles of Scilly. When it comes to coastal protection efforts that involve the local population, 54% of the respondents state that they do not personally participate in local flood and coastal protection as opposed to the 46% who do.

More insightful than the general figures on involvement in coastal protection are the figures disaggregated for the individual islands (Fig. 5.43). With the exception of Tresco, on the off-islands, more people are involved in coastal protection measures than on St Mary's. Involvement in coastal protection can take very different forms and is often related to the specific vulnerabilities of the respective islands. The most important features and types of collective coastal protection are described in the following sections.

5.3.2.1 Beach Cleaning and Site Clearance

Although beach cleaning and site clearance are not direct forms of coastal protection or climate change adaptation, they reflect the local communities' involvement in environmental management of coastal resources. This can be very relevant to adaptability towards environmental changes associated with the coastal zone, such as sea-level rise. Collective beach cleaning is practised on all the islands. On St Mary's, beach cleanings are usually organised by the Wildlife Trust and the

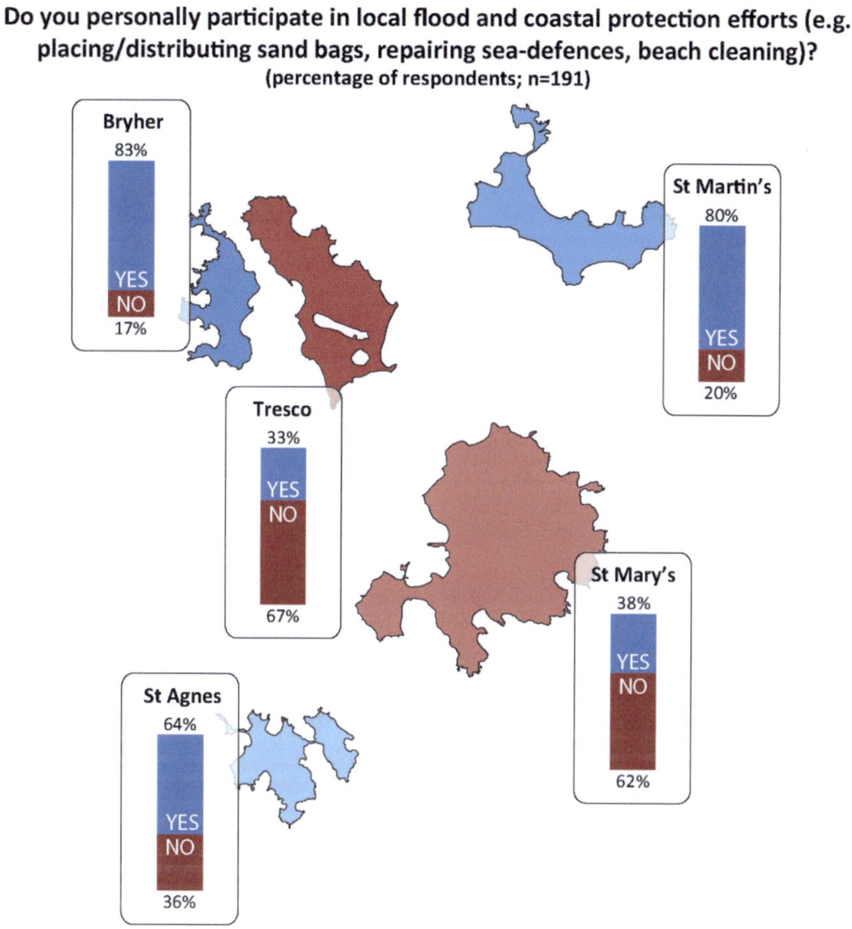

Fig. 5.43 Participation in coastal protection, per island

AONB, who provide bin bags and pickers. After storms, beach cleanings are organised not only by the Wildlife Trust or the AONB but also by engaged locals (field notes, 23/02/2014, 03/03/2014). Such initiatives are organised, for example, through online social media (field notes, 08/03/2014). Participation is especially high before the start of the tourism season in March/April (see Fig. 5.44). On the off-islands, the communities usually organise the beach cleanings themselves. On Tresco, however, the beach cleanings are initiated by the estate management and carried out by the staff (SL7: 59). Similar to community activity in general, collective beach cleanings are no longer as common on Bryher (SL1: 3).

The Coastal and Intertidal Zone Archaeological Network (CITiZAN) trains volunteers from the Wildlife Trust and the local Archaeological Group to monitor changes and erosion of the high number of coastal archaeological sites on the

Fig. 5.44 Beach cleaning and site clearance on St Mary's: Porth Mellon, 2014 (*left*) and Porth Minnick, 1989 (*right*, *source* Read 1990c)

islands. The continuing erosion of such sites close to the shore, however, seems inevitable because of sea-level rise and the low priority for coastal protection measures by the authorities (SP3: 124).

5.3.2.2 Coastal Protection/Maintenance of Sea Defences

On St Mary's, the most relevant sea defences are seawalls, which are maintained under the auspices of the council. There are a few individual properties, which have to be protected from flooding and storm damage by self-help measures, but the role of collective action is not as relevant here as it is on the off-islands, where 'it is vital' (SP3: 12). Some of the off-islands (e.g., St Agnes and Bryher) have seawalls that were constructed by the Environment Agency in the past. Professional maintenance and repair are rarely carried out, however. The Duchy of Cornwall as the landowner 'will have a watching brief, but the local people will actually do the work themselves' (SP16: 36).

On St Agnes, seawalls were severely affected by the storms in the winter of 2013/2014, and 'particularly after the last round of flooding […] the people were actually clearing up; the community were clearing up all the debris afterwards. They were doing the kind of essential repair work to some of the coastal defences' (SP10: 36). Signs of heavy erosion called for action because further storms would have probably caused the sea defences to be breached (SP3: 12). Local farmers and residents formed 'work parties' (SP5: 7) to fill the fragile areas of the sea defences with bags of rocks (see Fig. 5.45) and did so in a community effort (SC2: 2). They used the equipment of a local farmer, 'and everybody who had a vehicle came down to help' (SC5: 77). Locals also monitor coastline erosion. In the long term, external funding is essential to preserving the sea defences. One islander observes, however, that to a certain degree:

> there has to be a self-reliance […] – the people on St Agnes get organised and they go around and check all the defences and try to see what they can do. They will need outside help, but if they do the initial taking account of what they need, […] they have to do that, they cannot rely on other people to do that for them. (SP3: 78).

Fig. 5.45 Coastal protection on St Agnes: repaired seawall (*left*), and local solution to prevent erosion (*right*), both February 2014

On Bryher, dunes were overtopped and eroded and had to be stabilised with local materials, such as branches and large rocks, but partly also with garbage or scrap metal, by people from the community (field notes, 06/03/2014). In addition, a lead that was blocked, preventing floodwater from flowing back into the sea, had to be repaired by locals. This was 'more sort of patching up, afterwards. And there are some things that require long-term solutions, big solutions' (SP15: 92). In the past, but still, to a lesser extent, today, islanders also took responsibility for coastal protection themselves, e.g., by planting marram grass or using hedge cuttings to build up dunes (SC2: 27, SL6: 26). Such small actions have had a big impact in terms of protection for Bryher, where it 'has made a tremendous difference. It is almost impossible to describe to people what it was like before it was there' (SL6: 24). A practice that lead to conflict was the planting of the Hottentot fig (*Carpobrotus edulis*) (Fig. 5.46) to stabilise dunes from further erosion. The plant is considered an invasive species by the conservation organisations. Nonetheless, the resulting dunes protect many coastal areas today, where construction of seawalls would have been too expensive (SP12: 38).

On Tresco, most of the maintenance of seawalls or coastal protection is organised by the estate management, benefitting from its staff and the wide range of services, such as the forestry department, which possesses machinery and material that can be used to repair or build up damaged structures, such as the telephone cable, after the February 2014 storms (Fig. 5.47) (SP18: 13). Accordingly, as reflected by the household survey people are less involved in coastal protection measures on Tresco.

Fig. 5.46 Hottentot fig on Bryher, July 2014

Fig. 5.47 Reinforced telephone cable junction on Tresco, February 2014

5.3.2.3 Repairing of Coastal Infrastructure

On St Mary's, the council organises a direct labour force to monitor and check coastal infrastructure. Nevertheless, they have always relied on the community 'to feed back into them with information' (SP3: 80) and to help to clean up roads and

Fig. 5.48 Damaged slipway after storms (*left*), March 2014; the same slipway repaired in July 2014 (*right*)

Fig. 5.49 Slipway on St Martin's: locals discussing how to repair damaged slipway, March 2014

beaches after storms (SP10: 40), 'to restore Scilly's beauty' (Read 1990c: 19). This happened frequently after the February 2014 storms and the authorities benefitted from locals who monitored storm damage around the islands (Hargreaves 2014b). Particularly on the off-islands 'self-help is the order of the day', because 'people need to be independent and be able to do things. And if we did not do them, probably nobody else would' (SP14: 19), e.g., on St Agnes, where in the past a storm caused the electricity supply cable to be washed out and there was no quick assistance from the mainland or even St Mary's in repairing the damage. Similarly,

Fig. 5.50 Locals repairing St Agnes Quay after storm damage in 1998 (*Source* both, Mumford 1998: 9)

Fig. 5.51 Emergency response with sand bags in Old Town, 1990 (*Source* both, Gibson 1990: 45)

damages to the only quay, the 'most vital structure of St Agnes' (Mumford 1998: 7), often have to be repaired by the islanders themselves (Fig. 5.50).

On St Martin's, a slipway for gig boats and fishing boats was completely washed away. To launch their boats for the season, locals had to come together and repair the slipway using local material and machinery (Fig. 5.48). Members of the community gathered to discuss options and the best solutions for repairing the slipway (field notes, 05/03/2014; Fig. 5.49) and ultimately it was rebuilt by the islanders themselves (SL2: 5; SP12: 36). Also the Old Quay, which is hardly used today but has value as a landmark, was severely damaged by storms in the past, particularly last winter, and repaired by islanders with funds collected among community members (SP16: 32). On St Mary's, a pub was flooded and a café was damaged by the strong storm on 14 February 2014. Locals collectively installed provisional protection to avoid further damage to the building (Fig. 5.52). A person involved in the effort describes it as 'amazing – it was completely damaged one day, and within very short time everybody pulled together and put it back together again' (SP8: 102).

In response to the damages of the storms in the winter of 2013/2014 and their impacts not only on the Isles of Scilly but in the whole Southwest, the British

Fig. 5.52 Emergency response at café in Hugh Town after storms (*Source* both, A. Martin, private photographs, 14–15/02/2014)

government established schemes to assist local businesses and people who were affected. According to the council, however, little support was requested by islanders, probably because of the high degree of bureaucracy involved and people being busy preparing for the upcoming tourist season instead of waiting for money to come (field notes, 25/07/2014).

5.3.2.4 Emergency Response

Emergency response, especially on St Mary's, is coordinated by the council through an emergency responders forum, including the fire brigade, police, lifeguards, harbour master, hospital, health centre and the airport. In cases of emergency, the direct labour force 'goes out and puts sandbags out and puts the storm gates on slipways' (SP1: 8) and volunteers help to set up rest centres and care for people, including on the off-islands (SP17: 19). In the past, events also called for volunteers to retain sea defences during heavy seas, such as in 1990 (Fig. 5.51), or 1962, when 'a small army of islanders started to build up breached seawalls with rocks and sandbags [and] raced against time to effect repairs before high tide' (Mumford 1962: 27). Moreover, the council is working with the Environment Agency and MetOffice to provide heavy weather warnings and is involved in the local resilience forum for South West England for exchange with councils from other districts regarding flood protection (SP19: 30). The councillors fulfil the role of informing and helping people on the individual islands, as well as finding local solutions together with the affected people, e.g., the farmers (SP3: 14).

As mentioned above, cooperative disaster response is relevant and has been especially important on St Mary's, involving different local actors. As one of the islanders explains,

> if there was a really terrible situation, where there was risk to life, or a huge break down in defences, I am sure that the Duchy, Wildlife Trust, and the Council, and everybody would come together, because, well, they would have to. (SP3: 98).

Also another respondent suspects 'that we might not actually have any help from the mainland because they would be too busy in mopping up their own disaster' (SP1: 26). If evacuations were to become necessary, the council relies on households to take in evacuees because of the limited capacity of provisional emergency rest centres. Thus, apart from council officers and emergency responders, 'the community are actually key to these sorts of things' (SP10: 44). Many members of the community help each other with filling and distributing sandbags provided by the council (SP8: 98; Mumford 2008b), as well as preventing further damage to properties and shops.

Furthermore, people help each other in times of emergency, e.g., by cooking 'meals for people, [...] if they found that somebody had had a flood and their kitchen was flooded. [...] I think as a community we are pretty good at that' (SP8: 106). In times of emergencies, the smallness of the community helps, according to one of the islanders,

> because we can just get a hold of people, just like that. And we know, if they are not there, we know that they have gone on holiday or out, and we phone them. So, it is a lot easier here. Also, here, we are a lot better off, because everybody knows everybody, neighbours look out for the people next door, the vulnerable people that they know are vulnerable (SP17: 12; cf. SP1: 60).

These features are even more important on the off-islands, which, during very rough weather can become isolated from St Mary's and assistance from there or from the other islands is unavailable (SP17: 12).

In the household survey, most respondents (48%) claimed to act 'on their own or with family members' to protect their property and belongings during flooding and storms. Almost a third (29%) act 'self-organised with other members of the community' and the rest (28%) take no personal action (Fig. 5.53). However, most of the respondents who are not active personally are from Tresco and St Mary's (Fig. 5.54), confirming the observations and interview statements that, on those islands, the council and Tresco Estate, respectively, organise most activities. In the case of personal action, generally, individual action or action with family members is the most common form, while community action is also important on all of the islands, with the exception of Bryher.

Another form of emergency response is the example of help in case of ship accidents when people on the Isles of Scilly help to rescue capsized boats or ships in danger. Traditionally, that was one of the roles of the gig boats; a tradition that is retained through the sporting competitions. Today, locals are still the first to help boats in danger even before the official lifeboat reaches the place of emergency (SP17: 21). Thus, forms of collective emergency response are part of island life and have existed long before the discussion of climate change and sea-level rise.

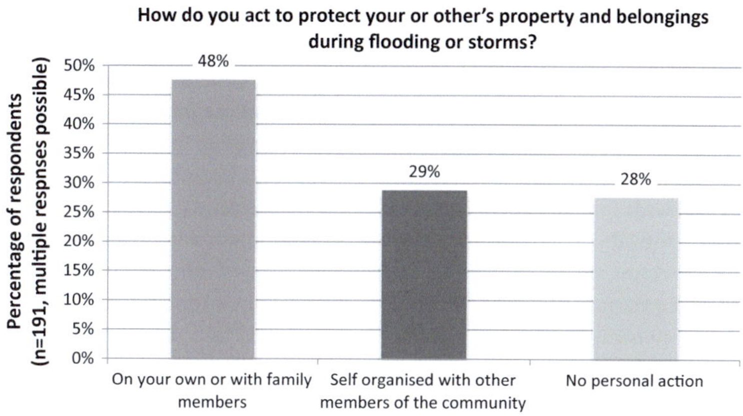

Fig. 5.53 Protection of property and belongings during flooding and storms

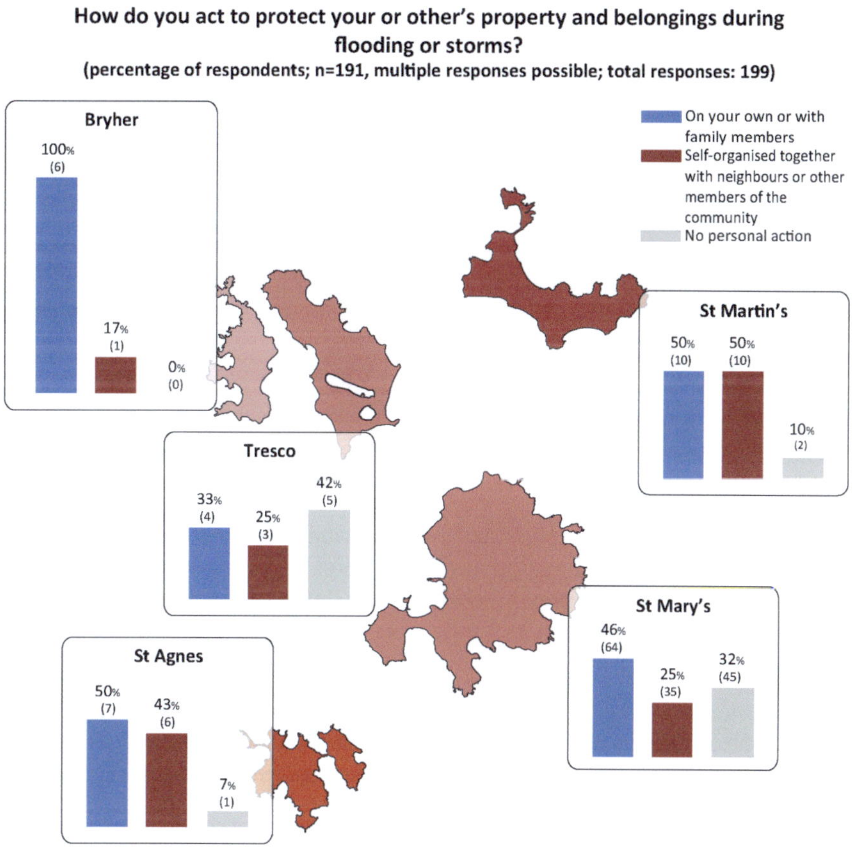

Fig. 5.54 Protection of property and belongings during flooding and storms per island

5.3.2.5 Climate Change Mitigation

In terms of climate change mitigation measures, generally, renewable energy technologies play an important role. Locals try to promote individual and collective projects, e.g., through Transition Scilly, for photovoltaic power generation, but, again, conflicting interests with the preservation and conservation elements have prevented a widespread and even individual development of such projects (SC6: 69). In addition, the financial aspect plays an important role. The council tries to facilitate investments in renewable energy generation and to use EU funding schemes and external investors for projects such as smart grids, waste-to-energy or wave power generation (SC1: 8). Because the islands 'are so separate' they could work as a 'laboratory to try out new technologies' (SP11: 42) thus generating new employment and income sources.

5.4 Findings from the Case Study

The preceding chapters have described the historical development of the Scillonian society, the islands' vulnerability to impacts from climate change, with a focus on sea-level rise and coastal flooding, as well as features of social capital, collective action and climate change adaptation today. This firm basis of data allows the questions to the islands regarding the case study (cf. Sect. 3.4) to be answered:

- Which elements of social capital are most relevant to climate change adaptation on small islands and what purpose(s) do these elements serve best?
- What are limiting and promoting factors for the mobilisation and effectiveness of social capital for climate change adaptation?
- To what extent do island-specific features play a role in social capital mobilisation for climate change adaptation? What are these features?

One approach to answering the questions is to test the hypotheses phrased in Sect. 4.1.1.

- [H1] Collective social capital increases adaptability on small islands by facilitating civic engagement and public participation in decision-making processes.
- [H2] Collective social capital increases adaptability on small islands by facilitating flows of information and increasing levels of acceptance towards adaptation measures.

Yet particularly the qualitative methods facilitated a focus on the differences between the islands, which helped to explain the relationship between specific community structures, contexts, vulnerabilities and their roles for adaptability. Can these hypotheses be validated or not? Upon which drivers and barriers does adaptability depend?

On the one hand, a verification of the hypotheses includes the analysis of the micro level (cognitive and structural domain) and macro level (legal framing, participation in decision-making and the role of political institutions) of social capital, with regard to climate change adaptation (Sects. 5.4.1 and 5.4.2). On the other hand, explanatory factors include locally specific features of a community, such as historical development and demographic structure (Sect. 5.4.2), and particularly the insular setting (Sect. 5.4.3).

5.4.1 Elements of Social Capital for Adaptation

As shown in Sect. 5.2, there is a high degree of social capital on all the islands, both in terms of the cognitive and structural domain. Section 5.3 gave an overview of how social capital can be relevant to climate change adaptation. On this basis, I can identify in the context of the Isles of Scilly the different types of social capital (bonding, networking and linking social capital), elements (e.g., trust, attitudes, participation) and the extent to which these elements are already or potentially relevant to climate change adaptation, incidentally as well as purposefully.

In some cases, long-term adaptation by coastal protection is not feasible, e.g., in the case of the low-lying areas on St Agnes (SC4: 20) or St Mary's, or even Hugh Town itself, where seawalls can even lead to further erosion. Also for extreme events, the adaptability of the islands might be too low in the long term; in case of 'an emergency that we cannot deal with ourselves' (SP17: 4), the Isles of Scilly 'rely heavily' (SP17: 4) on support from the mainland (e.g., Cornwall). However, the previous chapter described several examples in which the community plays an important role in adapting to risks from climate change.

The stakeholder and expert interviews I conducted with decision-makers and officials from the local authorities revealed how important a cohesive community, high levels of trust and participation of the population are for the Isles of Scilly to be able to adapt to climate change. To identify the elements of social capital most relevant to climate change adaptation on the Isles of Scilly, I will disaggregate the examples from the previous subchapter into the elements and types of social capital that can be identified, and determine specific features of adaptation to which they are relevant.

5.4.1.1 Cognitive Domain: Trust, Norms and Attitudes

For certain features of adaptation to climate change, bonding social capital (e.g., relationships between family members and friends) is particularly relevant. Bonding social capital is the basis for relations of trust, which become important for the islanders in times of emergencies, such as storm surges. People help each other to protect their houses from flooding and storm damages and are willing to accommodate others in case of evacuations. The local council relies on this high degree of bonding social capital and the cognitive elements of trust, solidarity and reciprocity within the local community. Volunteers can be recruited in a very informal way for disaster response efforts and information spreads quickly throughout the population. This compensates for the lack of resources of the small council and, therefore, contributes to the community's adaptability. These effects of social capital can be classed as incidental because, though they are not intended for emergency response, they are nonetheless indirectly very beneficial. Such effects can become purposeful, when organisations such as the council explicitly make use of family networks or neighbourhoods or encourage people to come together in public spaces to increase the flow of information and trust, explicitly in the context of risk management or adaptation.

An element of the cognitive domain of social capital is norms of reciprocity. On the Isles of Scilly, such reciprocal behaviour helps climate change adaptation, for example, in the form of people lending each other tools, material or even labour among non-governmental organisations. In this way, necessary resources (e.g., for coastal protection works and repairs), are available to people who otherwise would not have access to them, i.e., people with lower individual adaptability. Such norms of reciprocity develop informally and as purposeful norms, though not specifically for climate change adaptation. Organisations and clubs that facilitate networking social capital can promote them. Examples on the Isles of Scilly are the Farmers' Initiative, the local council and the Wildlife Trust, who bring together people from different business sectors to increase common adaptability. Such networking social capital facilitates not only sharing tools or ideas but also maintaining and renewing local knowledge and skills, such as farming practices. Examples that can be linked to climate change adaptation are traditional hedge cutting and stabilisation of dunes by planting certain plant species, or techniques of land management, such as using salt resistant crops and special soils. The local knowledge shared through such networks does not necessarily have to be traditional knowledge, but can also be newly developed techniques by locals. Again, these effects are not necessarily purposeful social capital, but they can be made purposeful for climate change adaptation.

Social cohesion, here operationalised by a high degree of bonding social capital and levels of trust, also results in a strong common identity as islanders and as Scillonians. This identity is strongly related to the natural environment and is accompanied with certain common values and attitudes that can contribute to adaptability. A very distinct value among islanders is natural protection and the protection of the islands' cultural landscape. The resulting attitudes and behaviour help explain environmental concerns and facilitate collective action. Yet social cohesion can also promote adaptability in another way: by facilitating compromise and the acceptance of change, if it is for the benefit of the whole community. This can be observed on St Agnes, where on local said:

> all of us are happy to live here, and the majority will accept the changes as they come, with a degree of kicking and screaming that we do not want to see change (laughing). But I think accepting the inevitable is also going to be accepted as part of the overall good of the community. (SL5: 30).

Additionally, however, there is a need for certain individuals to communicate and moderate changes so that 'the majority will accept change and the more active members of the community will understand the changes'. Moreover, there have to be people who are 'able to argue the case with the incoming, with those who maybe were less active, less prepared to accept' (SL5: 32).

5.4.1.2 Structural Domain: Collective Action and Decision-Making

An important feature of how social capital can promote adaptability towards climate change is by direct collective action by the population. Interviews and participant observation showed that self-organised activities are important, complementary to action by local authorities or other organisations, or as a substitute for lacking official initiatives. On the Isles of Scilly, there are several examples of collective action that directly refer to coastal protection and have been relevant in the past and present. Especially on the off-islands, collective action is crucial and people are motivated to be involved in coastal protection. Practices of collective action are part of the structural domain of social capital though influenced by the cognitive elements, as described in some of the previous examples. However, purposeful collective action initiatives for climate change adaptation or coastal protection are organised formally (e.g., by NGOs), as well as informally by locals within the individual island communities. Such initiatives become relevant to different strategies of shoreline management, as outlined in the SMP (cf. Sect. 5.3.1). The case of NAI implies no action by the authorities. Thus, if agricultural land or other areas that are not the priority of protection are affected by erosion or flooding, purposeful collective action is necessary. Similarly, managed retreat or realignment of coastal defences is facilitated by cooperation and support of the local population. Collective action in the context of adaptation also helps to reinforce social capital itself. The coming together of the community 'was one of the good things that came out […] of the storm damage this year [2014]' (SP10: 40), and, therefore, contributes to community spirit and the cognitive domain of social capital.

In terms of adaptation planning, social capital has limited potential on the islands. It is rather networking social capital that becomes relevant when people want to influence decision-making. Examples of this are campaigning groups involving people from across the islands who otherwise do not work together. Such purposeful networks are relevant on the islands since there is no formal involvement of the community in terms of adaptation planning. Networking social capital that has a fairly incidental effect for climate change adaptation is the feeding back of information about weather conditions, erosion and other coastal or environmental issues to the local council, e.g., by farmers, divers or boatmen.

In addition to bonding and networking social capital, linking social capital is also beneficial for the communities on the Isles of Scilly with regard to climate change adaptation. Linking social capital (the relationships and networks with people or groups outside one's own community, e.g., mainland UK) is provided by personal relationships, but also by participation in civil society groups. Such links can increase the islands' adaptability, e.g., through access to resources from larger parent organisations, as is the case for the Isles of Scilly Wildlife Trust. Connections to the mainland strengthen the communities' voices regarding policy-making. Moreover, such connections help to raise funds for local projects, such as natural conservation or renewable energy initiatives. Organisations such as Transition Scilly also contribute to awareness raising among the population about risks of climate change and opportunities of adaptation and mitigation. Apart from this purposeful role, there are also incidental effects of linking social capital. The Farmers' Initiative organises trips to the mainland or other islands to share knowledge and learn from other farmers about agricultural practices, as well to develop collective funding and buying schemes. This can contribute to farmers' adaptability to climate change, e.g., by helping them to deal with salt-water intrusion or unreliable transport due to rough weather conditions.

Linking social capital can become even more relevant in the future to fundraising for climate change adaptation because, as a person in charge of land management on the Isles of Scilly says:

> we have got to get together and work it out. We cannot just expect government handouts because I do not think it is going to come. But if you frame it in the right way, other funders might want to support us. And I am absolutely up for that. (SC4: 50).

The communities could develop strategies for coastal erosion management, which is usually funded by public authorities whose resources are available only to priority areas. Again, civil society groups are helpful for such projects and 'have to work in partnership in order to make the best use of resources. The money is sitting on the ground out there and we have got to work together to secure it' (SC4: 40). The local administration realised the potential of such networking and linking social capital. Thus, its objective is to 'instil the culture of collaboration, where we have not fighting or competing internally, but actually into competing externally' (SP11: 32).

Table 5.2 Summarised examples of effects social capital for adaptation on the Isles of Scilly

	Bonding social capital	Networking social capital	Linking social capital
Cognitive domain	• Informal recruitment of helpers • Quick spreading of information	• Norms of reciprocity—sharing of knowledge and tools for coastal protection • Facilitation of work of local civil society organisations	• Introduction of new/progressive idea, e.g., regarding climate change mitigation (renewable energies)
Structural domain	• Support of emergency services of the council • Mutual help in times of emergency	• Collective fundraising for community projects (renewable energies, coastal infrastructure) • Collective action for coastal protection	• Expert consultants and gaining of information for local projects (renewable energies, land management) • Fundraising from organisations and private people outside the islands

Table 5.2 summarises examples of the effects of cognitive and structural elements of social capital for adaptation to climate change on the Isles of Scilly arranged by type.

5.4.2 Drivers, Activators, Barriers

The previous section described the broad range of ways in which elements and different types of social capital contribute to the adaptability of the Isles of Scilly. However, there are also limits to the potential of social capital for adaptation on the Isles of Scilly (cf. Petzold 2016). The potential and limitations relate to drivers and obstacles for the mobilisation of social capital. I identified key factors, which each specifically influence the mobilisation and development of social capital for adaptation on the Isles of Scilly: land ownership and settlement structure, the role of institutions and civil society, clubs and informal groups, demographics and historical development, as well as individuals and leadership.

5.4.2.1 Settlement Structure and Land Ownership

A geographical feature that influences the mobilisation of social capital for collective action initiatives are the different sizes and settlement structures on the individual islands. While St Mary's is too large to be considered one community and collective action usually refers to different groups, on the off-islands they can refer to the whole population. There it is a matter of individuals who participate in community activities or not. This is an important issue to consider, because impacts of sea-level rise or storm surges, in this case, affect the whole community, while on St Mary's it is only a certain group of people that feels directly affected. However, whether the settlement structure (cf. Sect. 5.1.2), e.g., more scattered villages (as on St Martin's) or concentrated housing (as on St Agnes, and Bryher), has an impact on the mobilisation of social capital could not be definitively answered using the methods applied.

An important factor of how relations of trust, social networks, and civil society, lead to collective action, e.g. in land management or coastal protection, is land ownership. In this respect, the Isles of Scilly are a special case, having the Duchy of Cornwall as a single major landowner and the Wildlife Trust as an organisation in charge of land management. The limited number of freehold properties and area of public land consequently limits chances for people to be informally active in land management. This makes networking social capital, e.g., the Duchy Tenants Association, important, in influencing issues regarding land management, an issue that is especially important on the off-islands, in terms of waste management, freshwater supply and protection of properties and agricultural areas from coastal erosion and flooding. A special case is also the island of Tresco, where a private company runs a community. The consequence is that, whether with regard to land management, coastal protection or any kind of climate change adaptation, the estate management does not directly draw on high levels of structural social capital.

5.4.2.2 Role of Institutions and Civil Society

Related to the issue of land ownership is the role of institutions and civil society. Civil society organisations use and build upon existing social capital. By facilitating and organising collective action, they mobilise and at the same time promote social capital due to the tendency of social capital to grow when it is used (cf. Sect. 3.2.1). A positive example is the Farmers' Initiative, which brings farmers together and builds on networking as well as linking social capital, bringing in new ideas from the mainland, but also promoting traditional farming practices. In addition, they help in building trust by connecting farmers and promoting knowledge-sharing.

Organisations such as the Wildlife Trust are important for environmental education because many local skills are lost, creating a situation in which people are no longer able themselves to take care of their land. Civil society groups can return these skills to local people; 'as people become more apathetic, they probably need [such] bodies to remind them that this work needs to be done' (SP9: 75). This is an example of mobilisation of social capital. In the best case, such mobilisation is done in cooperation with locals who are already active; thus, it avoids disenfranchising those people who are actually skilled in taking care of the land themselves and might possess valuable local knowledge.

Generally, there are a few fairly dominant bodies on the Isles of Scilly (e.g., Tresco Estate, the Duchy of Cornwall, the Wildlife Trust and the Council of the Isles of Scilly) that are responsible for of land management and coastal protection, as mentioned above. These organisations bring in expertise and strategies from outside the islands, potentially undermining local practices and habits. The Duchy, on the one hand, 'almost [practised] social engineering' by controlling 'where people live' (SP11: 50) and suppressing community action. On the other hand, the Duchy also protects the islands from external influences such as over-development of touristic sites. The role of the council is also ambivalent because 'there is an expectation amongst the community now that the council will do everything' (SC4: 28) despite its structural limitations. Thus, the council can work as a facilitator of collective action, e.g., removing planning obstacles, and accessing funds (cf. SP11: 56), while it still relies on higher authorities. The Isles of Scilly Wildlife Trust, in addition to mobilising the community, can also hinder self-help measures such as planting marram grass or other plants to stabilise dunes, or using local materials for coastal protection.

Regulations, such as EU criteria for water management or construction (e.g., the island halls) can benefit climate change adaptation, but can also hinder informal collective action such as repairing quays and sea defences (field notes, 22/07/2014; SP1: 74). Water management on the Isles of Scilly is regulated under a special clause, which is why it is still operated by the council, unlike elsewhere in the United Kingdom, where water management is privatised (cf. Department for

Environment, Food & Rural Affairs 2014). EU regulations, however, increasingly influence local administration, forcing the local authority to adjust to EU standards. People report having the sense that 'outside influences are [...] going to override our ways of life' (SL5: 30). Whenever regulations are implemented in a top-down manner with experts being sent from organisations such as the Environment Agency, people see the experts as intruders who lack knowledge about the islands' special situation (SL2: 6). Moreover, the AONB and European Marine Site designations imply 'a very stringent and all-encompassing regulation' (SC1: 8), which can make it difficult to implement changes that have an impact on the natural environment. Accordingly, in terms of planning applications, conservation still has higher priority than coastal protection or climate change adaptation and mitigation measures (cf. SP1: 48, 52, 74). Ultimately, this could 'preclud[e] the development of the sustainable activities that could enhance and ensure that people survive on these islands' (SC1: 8); such 'activities' include renewable energy technologies, but also informal coastal protection measures.

To understand the relevance of social capital to the spread of information and acceptance of adaptation measures, it is important to review the institutional context: does involvement in the local council have any effect on these issues? Or are civil society and informal social relationships more relevant in this respect?

The inclusion of people in decision-making processes, e.g., by public consultation sessions and council meetings, can be important for any kind of policy planning, but also helps to keep people informed about issues that concern the community as well as the risks from sea-level rise and storm surges. In fact, the results of the household survey show that people who attend public council sessions and consultation meetings feel better informed about local affairs (Table 5.3). However, there is no significant correlation between attendance of such meetings and the perceived level of information regarding the impacts of sea-level rise or storm and flood events (Figs. 5.55 and 5.56).

On the other hand, while most people on the islands would accept further expenses in coastal protection measures, people who attend council sessions rate existing coastal protection measures higher on average than people who do not (Fig. 5.57).

Moreover, it is indirectly through civil society groups (which amongst others are organised under the umbrella of the AONB, also with regard to climate change issues) that people are integrated into decision-making. In the case of Scilly, such integration cannot be evaluated using quantitative measures because people are

Table 5.3 Correlation of attendance of council sessions with level of information

n = 191		Attendance of council sessions		Total (%)
		Yes (%)	No (%)	
Feeling informed about local affairs	Yes	93.8	76.6	84.5
	No	6.3	23.4	15.5
Total		100.0	100.0	100.0

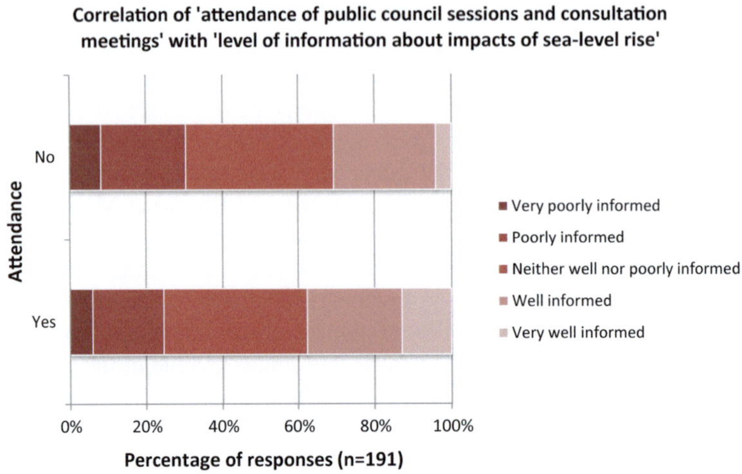

Fig. 5.55 Attendance of public council sessions and information about impacts of sea-level rise

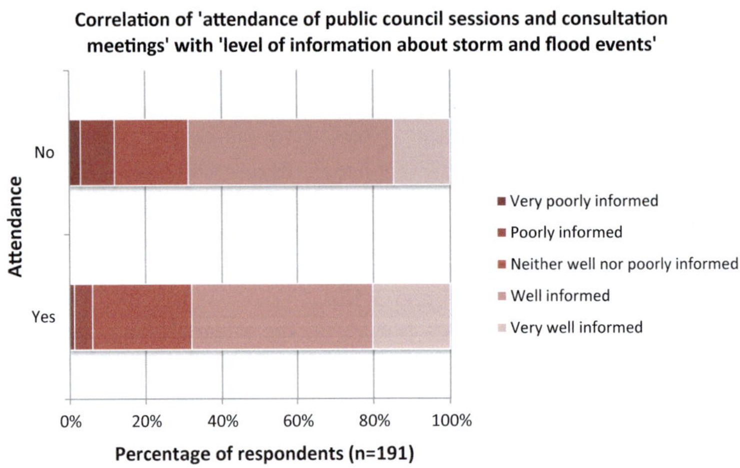

Fig. 5.56 Attendance of public council sessions and information about storm and flood events

often not formally involved with the groups while they still benefit from the groups' activities. This is an example of the collective understanding of social capital, as opposed to understanding social capital as an individual asset.

5.4.2.3 Clubs

Participation in social or sporting clubs and organisations is a typical indicator of social capital. Results indicated high participation in clubs, and some clubs, such as

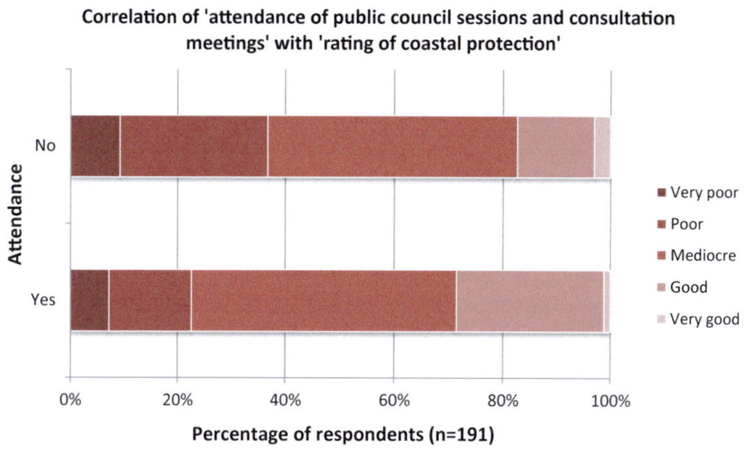

Fig. 5.57 Attendance of public council sessions and rating of coastal protection

gig rowing, play an important role in island life. The question is whether these clubs are relevant to the mobilisation of social capital in the context of climate change adaptation.

The survey results show that neither higher levels of activity in groups or clubs nor more frequent activities with friends or colleagues correlates with levels of information regarding sea-level rise or storms and flooding. Moreover, there is also no significant correlation between general group activity and participation in coastal protection (Table 5.4).

An example of the relevance of clubs, which cannot be reflected in the household survey, is their incidental effect of providing skills, such as gig rowing, or sailing, which can be important in emergency response, and which increase the islands' adaptability in heavy weather conditions. In fact, in the past, this was the one of the purposes of the gigs.

Thus, although clubs typically facilitate the increase of networking social capital, their role in community adaptability is not obvious. In the case of the Isles of Scilly, clubs certainly play an important role, e.g., in promoting community spirit, keeping up traditions, such as gig rowing, and building networking social capital.

Table 5.4 Correlation between group activity and participation in coastal protection

n = 191		Group activity			Total (%)
		At least once a week (%)	Less than once a week but at least once a month (%)	Less than once a month (%)	
Participation in coastal protection	Yes	46.6	51.9	44.0	47.1
	No	53.4	48.1	56.0	52.9
Total		100.0	100.0	100.0	100.0

Nevertheless, clubs are not the main driver of social capital mobilisation for climate change adaptation and are not the key factor for awareness-raising and facilitation of information flows on the islands.

5.4.2.4 Individuals and Businesses

The examples of collective action, an expression of social capital mobilisation, often do not depend on macro structures and structural elements, such as involvement in groups and clubs, but on certain individuals. These individuals can be people who are respected in the community and who initiate and organise actions such as beach cleanings. Partly, this is even more effective than when local authorities or organisations initiate such events (field notes, 03/03/2014). Individuals mobilise the community in the absence of institutions doing so, as on Bryher, where:

> they have got a couple of […] families that are the movers and shakers and make things happen. And again, you just need a couple of strong characters, who pull everybody together […]. On small islands, you are all going to do your bit. But you have got to have those people who make it happen. (SC5: 83).

On the other hand, collective action is often only possible if certain individuals possess the relevant knowledge, skills and tools to be able to take actions such as repairing seawalls. The availability of technology depends on individuals, such as machinery from farmers (e.g., on St Agnes, Bryher, St Martin's) or businesses (e.g., from Tresco Estate on Bryher/Tresco). For this reason, amongst others, the role of the farming community is still important, not only on the off-islands, because it can provide machinery for any purpose. The cognitive domain of social capital (mutual trust) is relevant to enable cooperation and sharing of resources between individuals who possess necessary skills and assets.

Individuals can also have a negative impact on climate change adaptation if they oppose such action, e.g., for personal reasons. On such a small scale as a small island, with a small authority, positions of leadership (e.g., in the Wildlife Trust or the Duchy) can have a very strong impact on any kind of development. In this sense, it can actually be individuals who block climate change adaptation or mitigation activities and obstruct the mobilisation of social capital.

5.4.2.5 Demographics and Economic Structure

Demographics is an important factor influencing how social capital is mobilised and which elements of social capital are more important. The Isles of Scilly have one of the oldest populations in the UK, which can be a problem when it comes to collective action that involves manual labour, such as participation in coastal protection efforts (see Fig. 5.58). In addition, the relatively small working population also implies fewer people available to provide the physical assets mentioned

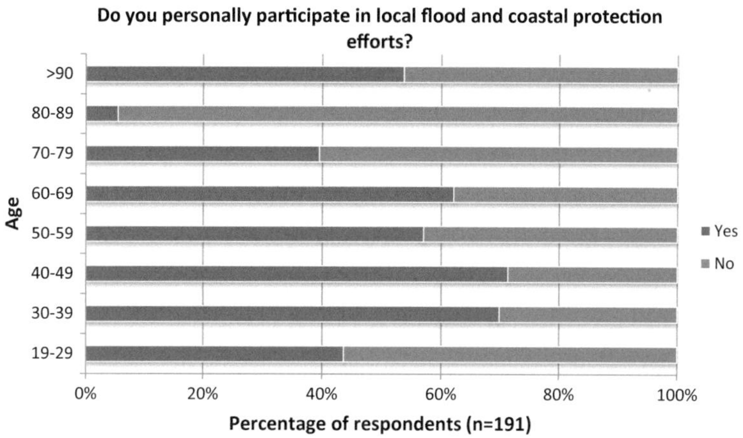

Fig. 5.58 Participation in coastal protection by age

above. In fact, the general ageing of the population and the decline in the number of native islanders is associated with the loss of traditional businesses such as farming and fishing.

The economic structure on the islands is also influenced in part by demographics. In addition to its relatively old population, the economy of the Isles of Scilly is also shaped by features such as a low unemployment rate, low economic diversity and a high self-employment ratio. Additionally, many people are forced to take up double employment due to the high rents and costs of living on the islands. There are multiple consequences of these economic circumstances.

A negative consequence is that people lack time to participate in volunteering for civil society organisations or to initiate collective action. Especially during tourist season, many people have two jobs or even more. Accordingly, day-to-day work is individual rather than collective, with the exception of St Agnes, where people are organised more collectively. On Bryher, the lack of interest and time for community action has even led to one of the island's two council seats remaining vacant for several years.

If an economy is diverse and more collectively organised, however, this can be a facilitating factor for social capital mobilisation that can also benefit climate change adaptation. A diverse island community can offer more skills and tools, as has proved important on St Agnes, where farmers and other businesses with tools came together to repair sea defences. On the other hand, people with a professional background on the mainland can contribute with networking skills and an academic perspective, which is important in representing the small communities on the off-islands with a strong voice at the local council and to the Duchy.

Moreover, the high self-employment ratio is also a factor that promotes individual self-reliance and independence. In fact, especially on the off-islands, people have proved to be very active and motivated to carry out community work, such as the roadworks on St Martin's and St Agnes, but also the planting of marram grass for the stabilisation of dunes. The example of Tresco, again, shows the opposite situation. The estate employs the majority of the population, and the management of the estate is in charge of everything. Individuals rarely push forward self-help measures, initiatives of collective action or informal working cooperation. It must be said, however, that the estate management is very efficient and coastal protection and land management are successful; there is little reason for people to take these matters into their own hands.

Again, the difficult economic situation and specific demographic structures on the islands have negative and positive impacts on the mobilisation of social capital. This leads to the next section, which highlights specific insular features important to connecting the islands' social capital and adaptability.

5.4.3 *Insular Factors*

To understand how social capital is mobilised, and why the drivers and barriers have the effects described in the previous chapter, it is important to examine the specific insular features that the Isles of Scilly have in common with other small islands. These features are the small scale of a place-based community, small land area, isolation, a certain population that is experienced with local environmental features, and local knowledge and skills. Can these factors explain some of the results from the previous chapter that seem to oppose my hypotheses?

It is difficult to identify island-specific features as they overlap with less specific factors influencing the mobilisation of social capital for adaptation to sea-level rise. Nonetheless, on the basis of the empirical data collected, I try to demonstrate that the influence of specific insular circumstances on the respective islands cannot be neglected. In some respect, these circumstances may even be the key variable for explaining the shortcomings of social capital theory. Many of these findings resulted from questions arising after the household survey and are based on interviews with locals and participant observation.

5.4.3.1 Smallness and Isolation

The small scale of the community, 'everybody knowing everybody', and the resulting high degree of bonding social capital and trust substitutes for the role of bridging organisations, which are more relevant in larger communities, where it is more difficult for people to get together without such facilitators. This explains the seemingly low correlations of group activity with levels of information and participation in collective action related to coastal protection. Islanders do not rely on formal networks for such action. On the other hand, there is a low degree of day-to-day cooperation (businesses are rather individually organised) and there are even signs of social exclusion and stigmatisation, which can make it difficult to push forward progressive ideas unless they are shared by the majority (SP4: 32; field notes, 29/07/2014).

Furthermore, their isolated location has, both historically and in the present, forced the communities on the Isles of Scilly to be more self-reliant compared to mainland communities, as shown by empirical data raised from each of the various research methods applied. To conduct major infrastructure works, machinery and workforce has to be brought in from the mainland, which can take long, be delayed or simply restricted due to lack of accommodation, as occurred during the repairs of sea defences in 1965 (Mumford 1965). While, on the one hand, their small scale and isolation add to the islands' dependence on external assistance, they also make them less of a priority in comparison to mainland areas. Due to their historical experience of isolation and self-reliance:

> in many ways, the islands have always been very adaptable anyway. Because of where we are, because we do not have the support from local services that you would have on the mainland. We have to make do and mend. And over the years, we have sort of become quite adaptable and quite adapted to facing whatever nature throws at us. (SP1: 4).

In fact, the islands' shifts in economic focus in the past have demonstrated their adaptability to changes in the world market, economic realities, limited resources and rough weather conditions.

Especially the off-islands are self-reliant, as mentioned in the previous chapters, and reflected by higher participation in coastal protection (Fig. 5.59). As one islander explains, they are even:

> much more independent [...] than the community on St Mary's, which is a little bit spoon fed by the council. [...] [P]eople living on the off-islands do not quite have that luxury. They put a few things up a bit more themselves, they are a bit more hands-on. (SP10: 38).

Native islanders are more pro-active than incomers, who look to authorities to repair storm damage, also observed on Bryher (SL1: 1). Generally, people are aware of the need for external assistance for long-term adaptation, and, as an island, they have historically been well-connected to external actors. Nonetheless, the more self-reliant communities on the off-islands have a higher sense of not being a priority and, accordingly, a lower expectancy towards support from the national government (Table 5.5).

The insular setting, and the informal way of organisation, the effectiveness of bonding and networking social capital without formal networks, coupled with the high awareness of local environmental conditions, e.g., storm impacts and flood events, are examples of social capital mobilisation becoming relevant to features such as emergency response.

Many people trust in their own capacity to tackle challenges in the future as they have done in the past, resisting upcoming difficulties also due to the lack of an alternative (cf. SP14: 53). Thus, on the one hand, the experience with the climatic conditions makes the islands adaptable. On the other hand, people are 'very cautious about change' (SM3: 70), which can pose a threat to future planning if it impedes radical changes that might be necessary due to climate change. The

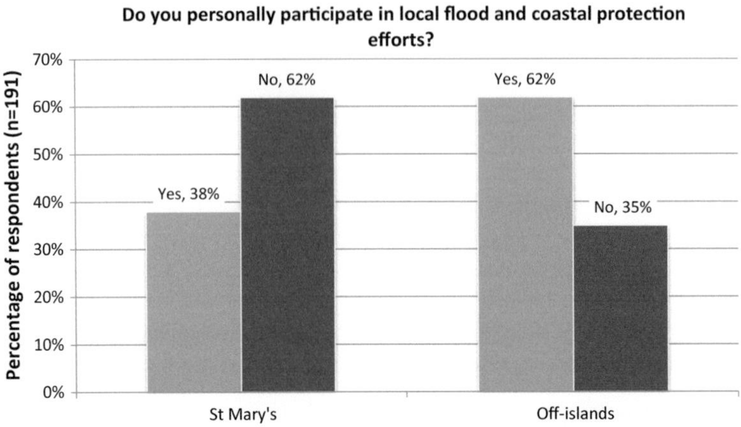

Fig. 5.59 Participation in coastal protection, on St Mary's and the Off-islands

Table 5.5 Expectation towards national government for support

n = 191		St Mary's (%)	Off-islands (%)	Total (%)
Support from government	Yes	74.2	54.9	68.9
	No	25.8	45.1	31.1
Total		100.0	100.0	100.0

isolation and very strong identity as being separate from Cornwall have created a 'philosophy of self-preservation' (SP6: 52), which has also led to scepticism and resistance to change, especially when it is initiated in cooperation with 'outsiders'.

Thus, one respondent thinks that the Isles of Scilly 'are not really ready for radical solutions to anything over here' (SC1: 18) because of general scepticism towards solutions from external experts (SP7: 14), but also the feedback from tourists, who see the islands as 'an idealised version of [...] the world as it was fifty years ago, where everything was simple, safe, unchanging' (SC1: 18). Accordingly, there is a 'negative perception of any activity that looks like manipulation of the environment or control' that produced a 'strong custom and local belief that everything should be left alone' (SC1: 18), a problem especially for conservation organisations such as the Wildlife Trust. However, this can also occur the other way around, e.g., regarding coastal protection and the planting of marram grass, considered an alien species (SP1: 70; SL3: 6).

5.4.3.2 Islanders and Incomers

An island-specific factor, which, however, can only be superficially covered in this book, is the role of people's family roots on the Isles of Scilly in contrast to people who moved to the islands from the mainland. Is there a different attitude, behaviour or even resilience recognisable between native islanders and incomers? Such a distinction is not obvious because it is unclear when exactly a person counts as an islander. Does one have to be born on the Isles of Scilly or is it enough to spend the greater part of your life on the islands? Or is it like one interviewee said: 'you cannot be a Scillonian unless you have got two grandmothers that were born here' (SP8: 8)?

The factor of having a family background and experience of living on the islands or not is important in terms of social capital because it can be very difficult for outsiders to be completely accepted on the islands. In fact, there are different levels, from tourists, to new resident, up to people who have been here for generations. This sometimes makes it difficult to find acceptance for solutions to certain problems. On the mainland, people often do not understand this problem (SP7: 13).

One distinction, an issue mentioned above, regards second-home owners. They might actually be native islanders, who, however, live outside the islands for most of the year. Their homes are let out to holidaymakers or they remain empty. Apart from contributing to the housing problem, these homeowners are absent from the islands most of the time, so they cannot contribute to everyday community life or to collective action, e.g., in times of storms or erosion. Second-home owners might, however, contribute to the islands by feeding back financial resources and bringing in connections to other mainland organisations and institutions, which the islands could benefit from, e.g., when it comes to the protection of archaeological sites or marine parks, or the improvement of transport links.

The main reasons for the argument about islanders and non-islanders, besides second-home owners, is the discussion around the 'helicopter generation', i.e., people who moved to the islands during the tourism boom and are supposedly more reliant on stable transport, e.g., by helicopter, as mentioned above. The argument is that people from the helicopter generation are less self-reliant, more active socially but resistant to change: '[i]n general, mainlanders seem to be more active, maybe like this they want to become part of the community. Locals sometimes do their own thing' (SM3: 56).

I received contradictory results regarding this hypothesis. While the argument was mentioned by several interview partners, the household survey does not reflect it. People who were born in Scilly are not less active than people who moved to the islands (Table 5.6). Similarly, this also does not apply to coastal protection efforts. Native islanders are not significantly more active in coastal protection than people who were not born in Scilly (Table 5.7). Thus, the argument that people who moved to the islands are more active in social clubs but less involved in the activities that increase the islands' adaptability cannot be supported by the survey result.

Further features regarding the differences between native islanders and incomers, which are difficult to assess with quantitative measures, are their attitudes and perception regarding environmental issues.

On the one hand, from my field stay and through interviews, I perceived that people from the helicopter generation are more concerned with sea-level rise and more alarmist, while they are less likely themselves to take action with regard to

Table 5.6 Correlation of origin on island and group activity

n = 191		Born in Scilly Yes (%)	Born in Scilly No (%)	Total (%)
Group activity	At least once a week	60.0	63.4	62.8
	Less than once a week but at least once a month	16.0	19.6	19.0
	Less than once a month	24.0	17.0	18.2
Total		100.0	100.0	100.0

Table 5.7 Correlation of origin on island and participation in coastal protection

n = 191		Born in Scilly Yes (%)	Born in Scilly No (%)	Total (%)
Participation in coastal protection	Yes	51.4	44.0	45.5
	No	48.6	56.0	54.5
Total		100.0	100.0	100.0

climate change. On the other hand, as shown above, non-islanders also claimed to participate in coastal protection while the survey results show only a slightly higher perception of risk (Table 5.8).

Clearer than the previous hypotheses about collective action and perception is the question of whether people would move to the mainland due to the risks of sea-level rise or coastal flooding. Not surprisingly, in this case, the readiness to move is higher among people who originally come from the mainland than among people who were born on the Isles of Scilly (Table 5.9).

While the helicopter generation theory cannot be statistically supported in this context, I identified different patterns that are more complicated than the helicopter generation argument suggests. My observations support the notion that people who were born on the islands or have spent a very long time here are less alarmed about sea-level rise and storms because they are accustomed to these phenomena. Newcomers, on the other hand, more often indicate that climate change is a significant risk. In terms of action, there is no meaningful difference between the two groups. Especially in terms of emergency response and collective action, such as beach cleanings, newcomers are as supportive as native islanders. Regarding more specific efforts at coastal protection, newcomers, however, seem more reliant on other islanders, who have the skills and experience in carrying out the works, while they themselves would otherwise have to wait for the council to become active.

Moreover, it is incomers who appeared to always have a certain vision for the islands' future, while locals are more pragmatic (field notes, 15/07/2014). In addition, while the helicopter generation might want to keep the islands as they are and have always been, and 'voice their concerns about it, because they see it [change] as unacceptable' (SP13: 20), I observed that among the more recent newcomers, there are certain progressive individuals who have more radical ideas

Table 5.8 Correlation of origin on island and perception of risk of sea-level rise

n = 191		Born in Scilly		Total (%)
		Yes (%)	No (%)	
Perception of risk of sea-level rise	No risk at all	0.0	0.7	0.6
	Very low risk	2.7	4.9	4.4
	Low risk	24.3	18.8	19.9
	High risk	64.9	61.1	61.9
	Very high risk	8.1	14.6	13.3
Total		100.0	100.0	100.0

Table 5.9 Readiness to relocate according to people's origin

n = 191		Born in Scilly		Total (%)
		Yes (%)	No (%)	
Readiness to relocate to mainland	Never	76.3	50.0	55.3
	Maybe	23.7	43.3	39.4
	Certainly	0.0%	6.7	5.3

of major change to deal with future challenges on the islands, including climate change (cf. SP1: 46). Initiatives to modernise and push forward development on the islands, whether by incomers or institutions, however, are met with strong opposition, as mentioned in the previous section.

5.4.3.3 Local Knowledge

A factor related to the previous one is island-specific local knowledge. Local knowledge regarding land management, e.g., practices of coastal protection or hedge cutting, is increasingly disappearing. This is mainly due to the islands' economic development dominated by the tourism industry. In fact, everything from boat timetables to shipping freights is organised according to tourism rather than supporting traditional trades, such as fishing and farming (SL6: 14). Traditional trades still play an important role on the off-islands only, and thus, there are still people with certain skills, as well as the motivation to apply self-help measures for coastal protection efforts. Organisations such as the Wildlife Trust fill the gap and take charge of land management, though, as mentioned above, not for coastal protection.

Local knowledge can be relevant to issues such as coastal protection and repairing of sea defences. Yet local knowledge does not necessarily involve skills and practices. Living on the islands is also about a certain everyday knowledge, an understanding of living with the consequences associated with conditions of isolation, such as unreliable transport, food shortages, storm impacts to infrastructure and flooding. Due to the insular situation and reliance on the weather, not only people who have jobs related to the environment are very aware of local weather conditions. Hence, people here do not rely on warnings from the Environment Agency as they do, for example, on the mainland (field notes, 10/03/2014). The attitude of self-reliance, feeling able to deal with difficult weather conditions and risks from sea-level rise is, therefore, not exclusive to islanders who were born on the islands, but rather it is an attitude that they particularly demonstrate while incomers can adapt to and learn from this attitude.

I would suggest that this more cognitive element of adaptability is a factor that cannot disappear so easily, because of the islands' isolation and small size which will always lead to a forced self-reliance. Practices and skills have been lost, however. They could be reanimated through efforts by civil society groups, the local council and informal networks. Thus, social capital is extremely important to promoting the conservation as well as the development of a specific island knowledge for climate change adaptation, which does not necessarily refer only to traditional skills and practices.

5.4.4 Summary of Empirical Findings

One major analytical method proved to be the comparison of the different islands. The five islands of the Isles of Scilly can illustrate the different elements, drivers and barriers, and island-specific factors for the mobilisation of social capital for adaptation to climate change. Many features are very ambiguous: do they support or hinder the mobilisation of social capital? Potentially, both. Therefore, it is important to understand exactly the mechanisms and circumstances of each island. From the empirical data collected, it became clear quite early in the research process that although the five islands composing the Isles of Scilly certainly have a common identity, they all have slightly different socio-economic and geographical realities (Fig. 5.60). Acknowledging and understanding this is an important step towards understanding the potential of social capital for adaptation on the islands.

On St Mary's, which has the most vulnerable settlement, social capital is most relevant in terms of collective emergency response. A driving factor is the high degree of bonding social capital, but also the integration of the community into response mechanisms by the council or initiatives by organisations such as the Wildlife Trust. On the other hand, informal initiatives of collective action for the coastal protection and land management are of minor relevance. Despite the high potential for community-based collective action also on St Mary's, there is a lack of actual initiative and 'speaking with one voice' to influence decision-making. As one resident says, many people 'applaud' (SL4: 3) but only few actually come and help.

St Martin's is less vulnerable to the impacts of climate change in the short term, due to its settlement structure and location. However, it is affected by storms and erosion, too. Social capital is high and there are relevant examples of collective action on the island. The demographic structure and fairly diverse community contribute to the island's relative self-reliance and adaptability.

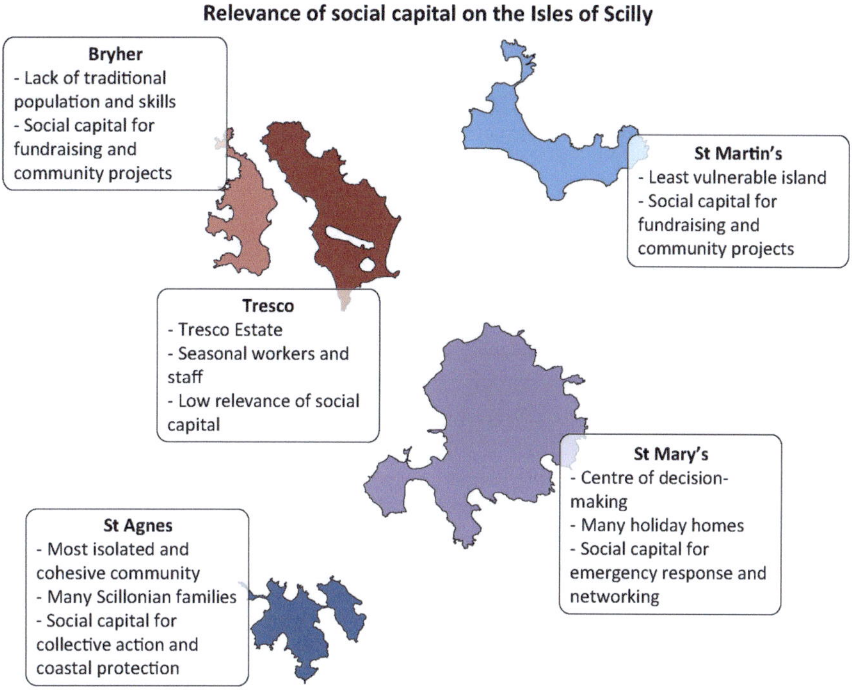

Fig. 5.60 Different relevance of social capital on the individual Isles of Scilly (cf. Petzold 2016)

Tresco is an example of an island where social capital plays no role in climate change adaptation. Although the island is vulnerable to erosion, flooding and storms in various places, the dominance of the estate management prevents informal activities from becoming relevant. Social cohesion, including seasonal staff, is mainly important for everyday community life. The resources and influence of the estate, however, could benefit the overall adaptation planning of the Isles of Scilly.

On Bryher, social capital could play an important role in adaptation as the island is vulnerable, but collective action is prevented by a lower degree of bonding social capital, a fairly competitive economic structure, the influence of Tresco Estate and the demographic shift through which there are hardly any native islanders left, who take the initiative of informal collective action.

St Agnes seems to be the most fitting example of community action in the context of climate change adaptation. It is the island that already suffers most from risks associated with climate change, but also the island with the highest degree of social capital mobilisation. Community action is relevant to coastal protection efforts, supported by locals with the necessary skills and tools, as well as by a very cooperative and diverse economic structure. Due to its relatively high isolation, the feature of self-reliance is most evident here.

The lack of verification of the hypotheses formulated in Sect. 4.1.1 directed the research towards a more explorative, explanatory perspective, which could also help to answer the research questions, supported by a more open and qualitative research approach. In summary, the hypotheses can be replied to as follows:

[H1] *Collective social capital increases adaptability on small islands by facilitating civic engagement and public participation in decision-making processes.* The results showed that people are formally involved in decision-making processes, but that structural social capital is not the influencing factor. Nonetheless, the high degree of bonding social capital on the islands is recognised by local authorities and non-governmental organisations as a potential for collective future development.

[H2] *Collective social capital increases adaptability on small islands by facilitating flows of information and increasing levels of acceptance towards adaptation measures.* While formal membership in clubs and organisations does not affect flows of information, the high degree of bonding social capital and smallness of the communities allow for a fast flow of information and a generally high awareness of issues and risks concerning environmental concerns. I found that levels of acceptance are closely related to attitudes of conservation and rejection of external consultants. In this case, strong bonding social capital rather than linking social capital can also lead to a lower acceptance of adaptation measures.

In summary, the analysis of the role of social capital with regard to island-specific factors contributes major insights, as these factors are shown to have a significant effect on drivers and barriers of mobilisation of social capital, both on the cognitive and structural level. On all the islands, the attitude of self-reliance, supported by the common identity, contributes to increased adaptability because people do not wait for actors from the mainland to help. While some findings suggest that this applies particularly to people who were born on the islands as opposed to incomers, this could not be definitively proven here. From a historical perspective, the structural element of social capital in the form of community life (e.g. in sports clubs, for fundraising or integration in decision-making), is probably even more developed today than in the past on Scilly. This does not necessarily increase adaptability because it does not relate to climate change adaptation. However, it illustrates a potential that has not existed in the past, by uniting people who, in the past, would never have united, including more people from the mainland who bring in different resources.

5.5 Reflection of Methods and Results

The variety of methods applied in this book allowed for an extensive analysis of the role of social capital for the adaptation to climate change on the Isles of Scilly. However, the research was also confronted with problems, while at the same time, certain methods proved more valuable than others. To provide a transparent account of my research, the following sections should critically reflect the research methods applied.

The two epistemological approaches and the different kinds of strategic methods applied were important for different kinds of results. The hypothetico-deductive method provided a basic understanding of social capital and people's perception of coastal risks, and how they deal with them. The grounded theory helped to explain some of the contradictions, because it provided the freedom to look in other directions, such as the historical development and the role of demographics. This information, mainly gained through stakeholder interviews, could partly be proven, and partly not, by means of the statistical analysis. Therefore, the ongoing theory development, which is suggested by grounded theory, relied in this case on the quantitative deductive approach I had designed, as well as the other way around: the mere statistical analysis would have left many key questions open in explaining the differences between the islands and the factors that influence the mobilisation of social capital.

In combination with the interviews, especially the participant observation and historical analysis of media and picture archives contributed to my understanding of island-specific factors. The participant observation also allowed for the inclusion of views from other locals than the ones whom I interviewed, as well as tourists, who have a different view on the islands. In fact, many tourists have been coming to the islands for decades already and have observed changes from an outside perspective, while at the same time they talk about issues locals would not, such as conflicts, taboos or simply issues locals are unaware of. Nonetheless, this group was only covered very superficially and my research lacks certain aspects that could not be covered with the methods applied. Also the general scepticism towards external 'experts' could have influenced how people talked to me and what they revealed to me about their community life. On the one hand, people confirmed that there is a general scepticism towards experts from the mainland. On the other hand, I did not perceive that people were sceptical towards my research. Instead, I found that people were very open and interested.

One circumstance of my research that could not have been planned beforehand but had a great impact in my whole research design were the storm events in the winter of 2013/2014, which partly coincided with my fieldwork. The events led to a stronger focus on the qualitative element because the household survey had not been designed to cover the perception of these events, though its results were certainly influenced by them. The coincidental timing of my research, moreover, certainly helped to facilitate dialogue about the topic of sea-level rise and climate change with locals. In addition, it afforded me the opportunity to observe some of the effects of the storms and flooding, which could increase with sea-level rise, and how people deal with these effects in practice.

The seasonality of my research also influenced the representation of certain groups. The main period of fieldwork took place in the winter. Thus, second-home owners, tourists and seasonal staff are not as represented as they would have been, had the main period of research occurred during the summer. This certainly applies to the survey, which was carried out in winter, while I tried to include other voices during the second major research stay in summer. Other factors in the winter period were that I was more easily recognisable as a researcher and people had more time and opportunity to talk to me than in summer when locals are particularly busy with tourism businesses.

Other issues that limited the interpretation, particularly of the quantitative data, were the small population on the off-islands, making it difficult to compare them with St Mary's or each other in a quantitative way, and the low external validity due to the focus on a small cross-case study. In turn, the results have high internal validity.

Furthermore, apart from the integration of a historical perspective as an analytical tool, my research resembles a snapshot. The islands are in a period of great structural change, with a new council administration and uncertain economic development. Are the islands moving towards a threshold? They might be close to one, not only in terms of viability due to climate change but also in terms of future pathways for the tourism and agricultural businesses, influenced by demographic changes. Organisations and stakeholders that I identified as relevant will soon disappear as new stakeholders emerge. Changes in the world market, international political sphere and national government can lead to sudden changes in priority and flow of resources to the islands. The demographic structure of small islands like the Isles of Scilly can change very quickly, with families developing and incomers taking over businesses.

References

Anon. (1987). A S(c)illy Attitude? In *Scilly Up to Date* (23): 2.

Araújo, I. B., & Pugh, D. T. (2008). Sea levels at Newlyn 1915–2005: Analysis of trends for future flooding risks. *Journal of Coastal Research, 24*(4A), 203–212.

Ash Futures. (2014a). Infrastructure Plan. Part of the strategic plan for the Isles of Scilly. Ashburton. http://www.scilly.gov.uk/sites/default/files/document/policy-documents/Infrastructure%20Plan%20final.pdf. Checked on 08/05/15.

Ash Futures. (2014b). Island Futures. A thriving, vibrant community rooted in nature, ready for change and excited about the future. A strategic economic plan for the Isles of Scilly. Ashburton. http://www.scilly.gov.uk/sites/default/files/events/Island%20Futures%20final.pdf. Checked on 08/05/14.

Ashbee, P. (1974). *Ancient Scilly from the first farmers to the early Christians. An introduction and survey*. Newton Abbot: David & Charles.

Bowley, R. L. (1980). *The fortunate Islands. The Story of the Isles of Scilly* (7th ed.). St Mary: Bowley.

Council of the Isles of Scilly. (2010). Climate Change Strategy 2010. Draft 1. http://www.scilly.gov.uk/environment/Climatechange.htm. Checked on 12/04/13.

Council of the Isles of Scilly. (2014). Helping Island Businesses Access Flood Funding. http://www.scilly.gov.uk/news/helping-island-businesses-access-flood-funding. Checked on 17/08/15.

Department for Environment, Food & Rural Affairs. (2014). Isles of Scilly Water Legislation Consultation. https://consult.defra.gov.uk/water/isles-of-scilly-water-legislation-consultation/supporting_documents/IoS%20Water%20Legislation%20Consultationformatted.docx. Checked on 23/01/16.

Dugan, E. (2008). The Isles of Scilly, where Britain's poorest people live. The independent. http://www.independent.co.uk/news/uk/this-britain/the-isles-of-scilly-where-britains-poorest-people-live-782992.html. Checked on 17/08/15.

FIFA World. (2011). The world's smallest football league. http://www.fifa.com/world-match-centre/news/newsid/154/345/6/index.html. Checked on 21/07/15.

Gibson, F. (Ed.). (1985). Eye Witness. The Isles of Scilly 1958–1984. St Ives.

Gibson, F. (Ed.). (1990). Eye Witness. The Isles of Scilly 1985–1990. St Ives.

Gibson, A. (1997a). A whole new variety of habitat. In *Scilly Up to Date* (103): 3.

Gibson, F. (Ed.). (1997b). Eye Witness. The Isles of Scilly 1991–1997. St Ives.

Grinsted, A., Jevrejeva, S., Riva, R., & Dahl-Jensen, D. (2015). Sea level rise projections for northern Europe under RCP8.5. *Climate Research, 64*(1), 15–23.

Grydehøj, A., & Hayward, P. (2014). Social and economic effects of spatial distribution in island communities: Comparing the Isles of Scilly and Isle of Wight, UK. *Journal of Marine and Island Cultures, 3*(1), 9–19.

Hargreaves, A. (2013a). Crime figures show Scilly is safest place to live in England. ScillyToday. http://www.scillytoday.com/2013/03/28/crime-figures-show-scilly-is-safest-place-to-live-in-england/. Checked on 21/07/15.

Hargreaves, A. (2013b). Island consortium chosen to clear Moorwell waste pile. ScillyToday. http://www.scillytoday.com/2013/11/28/island-consortium-chosen-to-clear-moorwell-waste-pile/. Checked on 10/07/15.

Hargreaves, A. (2013c). Scilly has lowest unemployment rate in the country. Scilly Today. http://www.scillytoday.com/2013/12/04/scilly-has-lowest-unemployment-rate-in-the-country/. Checked on 17/08/15.

Hargreaves, A. (2014a). Council launches project to get islanders driving electric cars. Scilly Today. http://www.scillytoday.com/2014/11/05/council-launches-project-to-get-islanders-driving-electric-cars/. Checked on 22/01/16.

Hargreaves, A. (2014b). Effects of winter storms on seabed revealed. Scilly Today. http://www.scillytoday.com/2014/06/26/effects-of-winter-storms-on-seabed-revealed/. Checked on 17/08/15.

Hargreaves, A. (2015a). Duchy undertaking improvement works on St Mary's and St Agnes. Scilly Today. http://www.scillytoday.com/2015/02/03/duchy-undertaking-improvement-works-on-st-marys-and-st-agnes/. Checked on 15/08/15.

Hargreaves, A. (2015b). Wave power as important as park house or infrastructure says chief exec. Scilly Today. http://www.scillytoday.com/2015/06/30/wave-power-as-important-as-park-house-or-infrastructure-says-chief-exec/. Checked on 17/08/15.

Hargreaves, A. (2015c). Wildlife Trust asks for vigilance over cattle fences. Scilly Today. http://www.scillytoday.com/2015/07/27/wildlife-trust-asks-for-vigilance-over-cattle-fences/. Checked on 08/04/15.

Heyworth, A., & Kidson, C. (1982). Sea-level changes in southwest England and Wales. *Proceedings of the Geologists' Association, 93*(1), 91–111.

Hickman, L. (2009). Isles of Scilly power down for E day. The Guardian. http://www.theguardian.com/environment/green-living-blog/2009/oct/05/isles-of-scilly-e-day. Checked on 08/12/15.

Isles of Scilly AONB Unit. (2010). Isles of Scilly. Area of outstanding natural beauty. Caring for Scilly's Exceptional Landscape, Supporting Sustainable Island Communities. AONB Management Strategy 2010–2014. Hugh Town. http://www.ios-aonb.info/wp-content/uploads/2012/09/IoS-AONB-STRATEGIC-PLAN-2010-2014-LR.pdf. Checked on 12/04/13.

Isles of Scilly Museum. ([around 1910]). Town Beach. Isles of Scilly Museum.

Isles of Scilly Museum. ([around 1960a]). Construction of Church Quay on Bryher. Isles of Scilly Museum.

Isles of Scilly Museum. ([around 1960b]). Construction of St Martin's Quay. Isles of Scilly Museum.

Isles of Scilly Museum. ([around 1960c]). Road works on St Martin's. Isles of Scilly Museum.

Isles of Scilly Museum. ([around 1960d]). Wooden sea wall. Isles of Scilly Museum.

Johns, C., Larn, R., & Tapper, B. P. (2004). Rapid Coastal Zone Assessment for the Isles of Scilly. A Report for English Heritage. Report No: 2004R030. Historic Environment Service. Truro. http://www.english-heritage.org.uk/publications/isles-of-scilly-rczas/islesofscilly20080116095450.pdf. Checked on 20/08/14.

Larn, R. (1994). *Shipwrecks of the Isles of Scilly* (2nd ed.). Nairn: Thomas & Lochar.
Matthews, G. F. (1960). The Isles of Scilly. A constitutional, economic and social survey of the development of an Island People from early times to 1900. London: George Ronald.
McCarthy, R. (2006). Ten years of tackling the Duchy. In *Scilly Now and Then* (8): 12.
Mortimer, T. A., Scourse, J. D., Ward, S. L., & Uehara, K. (2013). Simulated Late-glacial and Holocene relative sea-level and palaeotidal changes on the Isles of Scilly: a new approach for assessing changes in the areal extent of the inter-tidal zone. *Geoscience in South-West England, 13,* 152–158.
Mumford, C. (1974). Freakish tides and gales cause widespread havoc in Scilly. In *The Scillonian* (197): 12–15.
Mumford, C. (1985a). Scilly before the flood. In *The Scillonian* (221): 166–167.
Mumford, C. (1985b). What really happened to Scilly. In *The Scillonian* (222): 173.
Mumford, C. (1990). Islands ravaged by hurricane winds. In *The Scillonian* (231): 12–14.
Mumford, C. (1993). One large beach with a few boulders—and you're sinking! Scilly's massive coastal defence problems. In *The Scillonian* (236): 76–77.
Mumford, C. (1996). Spring start to £250,000 St Agnes coast defences. In *The Scillonian* (243): 76–77.
Mumford, C. (1998). New Year gales hit islands flower industry. Damage to St Agnes Quay. *The Scillonian, 247,* 7–10.
Mumford, C. (1999). Widespread weather damage. In *The Scillonian* (249): 64.
Mumford, C. (2000). Bank breach. In *The Scillonian* (251): 23.
Mumford, C. (2001). Comment: A doomsday scenario… In *The Scillonian* (253): 2.
Mumford, C. (2005). Heavy weather causes island travel chaos. In *The Scillonian* (260): 17–23.
Mumford, C. (2008a). Gales cause severe flower damage. In *The Scillonian* (267): 73.
Mumford, C. (2008b). Scilly battens down during fierce march storm. In *The Scillonian* (267): 102–103.
Mumford, C. (2009a). Trust replies to 'vandalism' charge. In *The Cornishman*, 2/26/2009.
Mumford, C. (2009b). Islands put £22,500 aside for fight against rising sea levels. In *The Cornishman*, 9/4/2009.
Mumford, C. (2010). Vulnerable Hugh Town flank is Porthcressa. Scilly Shoreline Plan looks to one-metre sea rise by 2100. In *The Scillonian* (271): 38–39.
Mumford, T. (1962). Heaviest seas in memory bring havoc to Scilly. In *The Scillonian* (149): 26–27.
Mumford, T. (1965). Major sea defence works on St Mary's, Isles of Scilly. Timber and boulders used. In *The Scillonian* (161): 6–7.
Office for National Statistics. (2011a). 2011 Census: Usual resident population by five year age group, wards in England and Wales. Table PP 04. http://www.ons.gov.uk/ons/rel/census/2011-census/population-and-household-estimates-for-wards-and-output-areas-in-england-and-wales/rft-table-pp04-2011-wards-in-england-and-wales.zip. Checked on 08/02/14.
Office for National Statistics. (2011b). 2011 Census: Usual resident population, local authorities in the United Kingdom. Table KS101UK. http://www.ons.gov.uk/ons/rel/census/2011-census/population-and-household-estimates-for-the-united-kingdom/rft-table-2-census-2011.xls. Checked on 21/10/13.
Office for National Statistics. (2013). Labour Market participation in England and Wales: Local Area analysis using the 2011 Census. http://www.ons.gov.uk/ons/rel/census/2011-census/detailed-characteristics-on-labour-market-for-2011-census-merged-wards-and-middle-layer-super-output-areas-in-england-and-wales/rpt-labour-market.html?format=print. Checked on 10/07/15.
Office for National Statistics. (2014). Self-employed workers in the UK—2014. http://www.ons.gov.uk/ons/dcp171776_374941.pdf. Checked on 10/07/15.
Permanent Service for Mean Sea Level. (2016a). PSMSL Data Explorer. St Marys, United Kingdom. http://www.psmsl.org/data/obtaining/map.html. Checked on 19/10/16.
Permanent Service for Mean Sea Level. (2016b). PSMSL Data Explorer. Newlyn, United Kingdom. http://www.psmsl.org/data/obtaining/map.html. Checked on 19/10/16.

Petzold, J. (2016). Limitations and opportunities of social capital for adaptation to climate change: A case study on the Isles of Scilly. *The Geographical Journal, 182*(2), 123–134.

Pound, D. (2004). Stakeholder dialogue. A good practice guide for users. Including 'Making the most of the Islands'—an example from the Isles of Scilly. Dialogue Matters. Wye, Ashford. http://www.dialoguematters.co.uk/downloads/goodpracticeguide.pdf. Checked on 21/08/14.

Read, J. (1985). Trustees to look after Scilly's unspoilt beauty. In *Scilly Up to Date* (5): 10–11.

Read, J. (1987). Serving the Community. In *Scilly Up to Date* (22): 10.

Read, J. (1988). Support for Scilly's Farmers: Essential Contribution to the Quality of the Islands. In *Scilly Up to Date* (32): 14.

Read, J. (1990a). I.O.S Museum. A History of Storms. In *Scilly Up to Date* (44): 11.

Read, J. (1990b). One night in December. In *Scilly Up to Date* (44): 19.

Read, J. (1990c). Volunteers came for Spring clean up at Porth Minnick. In *Scilly Up to Date* (45): 19.

Read, J. (1992). Anti-social activities. In *Scilly Up to Date* (67): 10.

Read, J. (1998a). A Generous Gift. In *Scilly Up to Date* (116): 2.

Read, J. (1998b). A Lucky Escape. In *Scilly Up to Date* (116): 1–2.

Royal Haskoning. (2010). Cornwall & Isles of Scilly Shoreline Management Plan. Revision 2. Isles of Scilly. Management Zones PDZ18, Management Areas 42–46. http://www.scilly.gov.uk/sites/default/files/document/planning/smp2.pdf. Checked on 13/08/15.

Sisam, K. (1962). Changes in sea level. *The Scillonian, 149,* 43–45.

Smith, J. (2008). Transition Scilly. In *Scilly Now and Then, 23*, 24.

Tangye, M. (2007). Some extremes of weather in 19th c Scilly. In *The Scillonian* (265): 163–166.

Thomas, C. (1985). *Exploration of a drowned landscape. Archaeology and history of the Isles of Scilly*. London: B.T. Batsford.

Tregarthen, C. (2000). Apparently we've got 200 years—and then some of us are sunk! *The Scillonian, 252,* 155–158.

Wakefield, H. (1986). A parable for our islands? In *Scilly Up to Date* (12): 9.

Ward, R. (1985). Boats for rescue, smuggling and racing. In *Scilly Up to Date* (1): 10–11.

Woodley, G. (1822). *A view of the present State of the Scilly Islands*. London.

Chapter 6
Social Capital as a Factor for Resilience

The data from the empirical research presented extensively in the previous chapter allowed for extensive insight into the case study, with diverse results. In this chapter, I retrace the theoretical concepts introduced and integrated in Chap. 3, adding to the development of the theoretical discussions.

First, I build on the results from the case study to refine and integrate the different theoretical elements of social capital, social-ecological resilience and nissology. Second, to answer the main research question of what shapes collective adaptability to climate change in island societies, I present a framework that links the different theoretical approaches specifically in a context of resilience. Third, I show limitations and issues not covered by the framework and my research approach.

6.1 Linking Social Capital and Social Emergence on Small Islands

With climate change as a social challenge, communities' responses become increasingly important (cf. Chap. 2). In addition to formal and static adaptation planning, informal relationships affect communities' adaptability because they produce emergent behaviour, which cannot be explained by analysing individual capacities alone (Sect. 3.1). Yet how these relationships develop and which effect they have for a community differs in each place and at every point in time. Equally different are the impacts and features of environmental changes and those associated with climate change and sea-level rise. In addition, the geographical context and changing environmental conditions influence the development of emergent behaviour and social structure exerts downward causation on individual and collective behaviour.

The basis for the framework includes bringing together the three strings of theory. In particular, my research approach consists of the integration of features of

J. Petzold, *Social Capital, Resilience and Adaptation on Small Islands*,
Climate Change Management, DOI 10.1007/978-3-319-52225-8_6

collective social capital with path dependence, social memory, non-linearity, emergents, downward causation, aspects of resilience and nissology, in a context of adaptation to sea-level rise on small islands.

6.1.1 Social Capital and Path Dependence

Path dependence shapes demographic and institutional structures (which can have long-term effects on the nature of social relationships) and the effectiveness of social capital for adaptation. The empirical case study of the Isles of Scilly revealed how path dependence and the composition of the demographic structure influence how effective social capital in the form of collective action can be, e.g., regarding adaptation to sea-level rise. As with other islands in a peripheral location, the Isles of Scilly show a certain trend of emigration and a demographic structure with a relative gap in productive workforce and specifically qualified people, i.e., human capital. In times of certain environmental changes, such as sea-level rise, when people are needed for collective action or for the sharing of certain skills, this demographic structure has a negative effect on the mobilisation of social capital and can be partly explained by path-dependence.

Singular historical events can shape how islands develop. The Isles of Scilly demonstrate that this perspective helps in understanding why individual islands develop differently in terms of social capital's relevance to adaptation. The change in land ownership, which was an independent decision by the landowner, led to an economic shift on the islands, implying a certain lock-in situation, including its economic but also demographic structure. Consequences were the disconnection of the community from land management and an influx of people from the mainland who did not have any experience with life on a small island. These are key factors that affect the relevance of the existing social capital for purposeful adaptation to climate change.

The example of Tresco, furthermore, illustrates path dependence by the last remainder of historical land ownership, which still has a strong influence today. Though only one of the five islands of the Isles of Scilly remains under the lease of the family that has been governing the islands for decades, their influence as the main economic player on the archipelago is still highly important. Tresco Estate provides the link to the most affluent clientele among the tourists (i.e., linking social capital), as well as financial and technical capital for structural adaptation measures. The latter, however, as has been demonstrated, also has a limiting effect on local social capital mobilisation.

6.1.2 Social Capital, Social Memory and Local Knowledge

Social memory is another issue where 'history matters' for the role of social capital for adaptation to climate change. Empirical analyses of Scilly show a collective social memory of having always been adaptable, leading to a certain feeling of self-reliance and independence, but also to stagnation. This social memory is promoted by collective self-help measures, both in terms of purposeful and incidental adaptation, as well as by literature, island-specific social groups and networks, and traditions such as the sporting and rowing traditions.

Closely related to social memory is local knowledge, here understood as attitudes and patterns of behaviour specific to life in local environmental conditions (cf. Mercer et al. 2012). Local knowledge is important on an individual level, but also in translating existing social capital into action, i.e., mobilisation of social capital. In terms of adaptability, in this case, local knowledge refers to traditional knowledge as well as knowledge and skills brought in by newcomers. Traditional knowledge includes land management or shared practices of disaster recovery and prevention, while the integration of incomers provides new knowledge, skills and ideas to the islands, as well as potentially new contacts through linking social capital. On the Isles of Scilly, such resources are provided by incomers who stay and settle on the islands, invest there and realise new projects, or by others, such as second-home owners or temporary staff, who regularly visit and provide financial resources and bring in skills such as administration, planning or land management, which can benefit climate change adaptation.

Thus, social capital has the potential to provide, maintain and increase local knowledge, both in a traditional and a progressive sense.

6.1.3 Social Capital and Non-Linearity

The issue of non-linearity must be considered in understanding social capital and its role for climate change adaptation. On the one hand, environmental changes and impacts associated with climate change occur in a non-linear way: rapid onset events such as storm surges, as well as slow-onset events such as sea-level rise, erosion and effects of changes in biodiversity. This fact implies that incidental social capital is as important as purposeful social capital in contributing to adaptability towards surprising events and thresholds (cf. Urry 2005: 7). The results from the case study show that both dimensions are relevant on small islands in order for communities to cope with events of emergency, where levels of trust and a strong bonding social capital have a significant incidental effect on mutual help and assistance. The structural elements of civil society and collective action can be considered to have incidental and purposeful effects for adaptability in terms of organised disaster response during unpredictable emergency events. Ultimately, uncertainty for islands such as the Isles of Scilly implies that the threshold at which the main settlement becomes uninhabitable is unknown, despite models and projections.

On the other hand, non-linearity also refers to social emergence, with individual relationships as its constituents. The way social relationships produce emergence, and the type of social relationships, as theorised by social capital, can depend on small and spontaneous events and coincidences, which links back to path-dependence. However, this refers accordingly to future developments. Regarding adaptability to climate change, any change in societal composition has an effect on many of the activators and barriers for the mobilisation of social capital such as demographics or leadership (cf. Sect. 5.4.2). Vice versa, unforeseen changes in land ownership, higher-level regulations or economic circumstances have implications for the mobilisation of social capital, as will be explained further below in terms of feedbacks and downward causation.

The findings from the case study show that small changes in demographic structure, which happened by chance, such as the choice of few native island families to move back to St Agnes, has led to the development of a highly cooperative and cohesive island community with a high degree of collective action that has already become explicitly relevant to adaptation to sea-level rise.

6.1.4 Social Capital, Stable and Ephemeral Emergents

The empirical analysis of the Isles of Scilly showed how social capital theory can add to the understanding of the development of stable and ephemeral emergents in the context of climate change on small islands. A high level of bonding social capital and levels of trust, norms of reciprocity, as well as a high relevance of informal collective action (on the structural domain of social capital) are crucial factors for the constitution of stable emergents, such as social memory and shared geohazard experience. This becomes evident in behavioural patterns in times of emergency and coastal protection, but also regarding the sense of place and perception as a self-reliant and adaptable island community.

At the same time, networking and linking social capital, which promote informal and formal collaboration, also lead to ephemeral emergents, which I identified by the work of civil society and the interaction of various institutions and organisations on the islands that deal with environmental management and climate change adaptation. As the two dimensions of emergents differ, the structural and cognitive domains of social capital have ambiguous relevance for adaptation to climate change and can even contradict each other. Empirical results from the Isles of Scilly, amongst others, showed that formal participation in social networks/clubs does not necessarily increase people's awareness of risk or level of information. High levels of trust and strong bonding social capital are important for emergency response but do not go along with cooperation and future planning.

6.1.5 Social Capital, Downward Causation and Agency

In an understanding of strong emergence, the concept of downward causation helps explain why certain patterns of behaviour on the individual level develop from existing social facts (Sawyer 2005: 72). This parallels the understanding of social capital as a result of individual interaction while at the same time social capital is a resource for the individual or excludes individuals. Thus, if social capital can be understood as the structure for emergence, then the barriers and drivers I identified (cf. Sect. 5.4.2) reflect feedbacks with downward causation. These, however, have to be evaluated differently for each research question, e.g., according to geographical context (e.g., insular factors, cf. Sect. 5.4.3). Here, certain factors can exert downward causes in a negative way for adaptation but in a positive way for conservation or other issues.

The 'dark side' of social capital, mentioned above, entails that, besides leading to exclusion, strong bonding ties and an emergent social structure with rather conservative norms and patterns of behaviour, social capital can limit adaptability towards climate change because it implies less openness to change that may be necessary. On the Isles of Scilly, the rejection of renewable energy systems or coastal engineering schemes developed outside the community (e.g., on the mainland) are examples of such downward causation of emergent social structures on individual behaviour, with negative effects.

While the concept of agency is important in understanding individual motivations for action and as a basis for emergence, the examples mentioned above demonstrate how emergent social facts, in turn, have a downward causation on individual agency and, thus, on the mobilisation of social capital. The epistemological consequence for social emergence is that methodological collectivism and individualism must be combined and that feedbacks and downward causation can never be defined in a universal or generalised way, but only on a case-by-case basis.

Another example of downward causation is the role of institutions such as the council, because 'there is an expectation amongst the community now that the council will do everything' (SC4: 28) despite its structural limitations. Responsibility is taken away from the community, resulting in a decline in agency caused by higher-level order.

Moreover, feedbacks from changing environmental conditions can have significant implications when they stimulate the mobilisation of social capital and, thus, adaptability, e.g., the coming together of the community on the Isles of Scilly, which 'was one of the good things that came out [...] of the storm damage this year [2014]' (SP10: 40).

6.1.6 Social Capital and Resilience

Besides identifying which types of social capital are relevant to adaptability and by which factors this is influenced, distinguishing which aspects of resilience are increased helps to place the theory in a wider social-ecological context (cf. Sect. 3.1.1). The four aspects of resilience are latitude, resistance, precariousness and panarchy (Berkes and Ross 2013).

On small islands such as the Isles of Scilly, issues of scale imply a low resilience, especially in terms of latitude (L), the number of changes a system can cope with before reaching a threshold. Economies are concentrating on one sector, here tourism, and adaptation strategies are often not very diverse because each stressor is significant enough to potentially push the islands' systems over a threshold. Here, sea-level rise is the focus, but obviously economic and demographic pressures add to the systems' low latitude. Linking and networking social capital are important in increasing adaptability towards diverse pressures because they add to the limited human and financial resources available in a small community. This increases resilience towards multiple stressors, which the islands face in addition to climate change adaptation. Bonding social capital, which refers mainly to relationships among family members and close friends, could not be enough to handle a diverse set of pressures, because a variety of skills and resources are relevant, which linking and networking social capital could provide for the affected community.

Resistance (R) refers to how resistant a system is towards changes, i.e., how difficult is it to alter the system's features, also depicted as steepness. Here, bonding social capital plays an important role, because it is associated with strong ties and high levels of trust. Moreover, stable emergents, such as collective behaviour and collective memory affect adaptability in terms of resistance, by strong bonding social capital and norms of reciprocity, as observed on each of the Isles of Scilly. This, however, can also have negative effects, e.g., when collective social memory produces conservative attitudes and rejection of measures which imply change. The empirical results from the case study showed that the community's feeling of self-reliance is accompanied by scepticism towards newcomers' radical ideas for change, adaptation and climate change mitigation on the islands. Therefore, a lack of linking social capital can increase resistance and potentially negatively influence adaptability.

Linking and networking social capital can be highly relevant to precariousness (Pr), the distance of a system to a threshold. The closer a system is to the threshold, the lower the resilience, i.e., the smaller the amount of change necessary for transition. Linking social capital, understood as relationships of more distant people and communities to each other, is important when radical solutions, such as relocation to another island or the mainland, are necessary for adaptation or even transformation, e.g., when islands become uninhabitable (at least for some) due to sea-level rise or when the economy collapses due to the lack of viability of tourism. Uncertainty and non-linearity, however, imply that precariousness is difficult to identify if thresholds cannot be defined.

A system's embeddedness in dynamics of change and adaptive cycles on lower and higher scales, described by panarchy (P), is closely connected to linking social capital, which can facilitate inflows of information and recognition of processes on other scales, as well as the capacity to influence processes on other scales. Thus, it is a two-fold issue, which in the case of the Isles of Scilly can refer to influence on NGOs based on the mainland (e.g., the Wildlife Trust or the RSPB, which are active on Scilly), but also decision-making regarding climate change adaptation policies nationally (e.g., central government in London) and internationally (e.g. EU), affecting the mobilisation of social capital on the community level.

6.1.7 Social Capital, Emergence and Islandness

The empirical research on the Isles of Scilly demonstrated the relevance of island-specific factors to adaptability to sea-level rise, by applying the concept of social emergence through a social capital lens. What can the specific study of small islands contribute here to the theoretical development of social adaptability?

In isolated areas, social capital is a key factor shaping society's self-organisation in terms of response to changing environmental conditions. Social capital, here, is already a precondition for survival. Thus, 'place matters'. Without high degrees of bonding but also linking social capital, island societies would struggle to keep up their social system, due to the various vulnerabilities. Islands and place-based communities that are accustomed to hazard-exposure develop stable emergents, e.g., a certain attitude and behaviour that is manifested in bonding social capital, high levels of trust and mutual help in times of emergency. Social memory (Wilson 2015) and geohazard experience (Manson 2001: 410; Ratter 2013) are particularly relevant features of small islands.

However, how this influences adaptability has to be analysed with regard to specific risks or issues and depends on many factors, as outlined in the empirical analysis of Scilly, where small local differences between the individual islands can lead to large discrepancies. Nonetheless, a common pattern could be identified, which matches other empirical findings. On the Isles of Scilly, it is the high degree of bonding social capital, which becomes mobilised purposefully for emergency response, or incidentally for spreading information and collective action. Formal social networks play a relatively lower role here in the development of stable emergents and the benefits associated with social capital, due to the pre-existing island context producing high bonding social capital.

6.1.8 Islands as Complex Systems

Studying islands 'on their own terms' (cf. Sect. 3.3.3) and as complex social-ecological systems (cf. Sect. 3.1.1) allows for the identification of epistemological and conceptual implications.

Seeing islands as special places due to their vulnerabilities neglects the complexities and, in fact, particular insular paradoxes of islands. On the one hand, they feature a certain shared social memory (e.g., self-reliance, resilience and geohazard experience), while on the other hand, they are characterised by interconnectedness and interdependencies, both internally (intra-island/-archipelago) and externally (island-continent, island-island).

The definition of the (island) system and its environment (cf. Sect. 3.1.1) is a point where research can go wrong before it even gets started. Here, the case study is composed of five separate islands, which at the same time form an archipelagic community. Especially on the largest island, St Mary's, there are various settlements which could be defined as relatively distinct communities. Archipelagic thinking implies considering these seeming contradictions by recognising on the one hand, e.g., a strong shared identity throughout the archipelago, which relates also to social emergence by collective memory and geohazard experience, shared language and names. On the other hand, place-specific conditions and path dependence have implications that lead to a completely different relevance of social capital on the various islands that comprise the archipelago. Each of the islands could be considered a small island due to issues of scale and location while the archipelago as a whole also features these vulnerabilities. This shows also that small systems are not less complex than large systems. Neither small size nor a small number of elements involved has a significant bearing on a system's complexity.

The empirical results from the Isles of Scilly show that systems and environments are a matter of epistemic intention and, at the same time, the system and its environment cannot be understood without each other. The features of social emergence that are relevant to adaptability in each islands' system refer to bonding, networking and linking social capital, which can each only be explained by the archipelagic constellation and the archipelago's relationship to the sea and to other continental regions.

The ecotone approach suggests a specific focus on the marine sphere of social-ecological systems. In fact, the water does not necessarily separate the islands, rather it connects them, e.g., boating is very common and part of many islands' identities and, in the present case study, at low tide one can even walk between some of the islands. This means that for the study of islands' adaptability the islanders' relationship with the sea, as well as inter- and intra-island networks have to be considered when analysing social emergence through a social capital lens. This perspective also helps by distinguishing bonding, networking and linking social capital, thus avoiding the misinterpretations of a constructed distinction between islanders and non-islanders, but rather focussing on the nature and implications of these relationships.

6.2 Framework for Social Adaptability to Environmental Change

The various issues described above show how the analysis of social capital helps explain why certain emergents, such as collective habits of land management and cooperative island businesses, can contribute to adaptability in some places (here, islands) more (e.g., St Agnes) than in others (e.g., Bryher, Tresco). While separately, social capital theory and social emergence already have high explanatory power, the empirical analysis of Scilly proves how a combination of the approaches provides them with crucial insights for understanding each other. Social capital theory adds to social emergence by distinguishing types and dimensions of social relationships and explaining respective implications for community resilience. Complexity theory and resilience frame social capital in a more dynamic understanding through their connections to concepts of non-linearity, path dependence and downward causation. The two approaches in combination, therefore, contribute to the understanding of micro-macro dynamics in social-ecological systems, as exemplified here for specified resilience towards sea-level rise on small islands.

Figure 6.1 combines the key features that influence adaptability on small islands from a perspective of social emergence through a social capital lens. It includes the factors of complexity that influence the emergence of higher-level order via stable and ephemeral emergents from lower level structures, i.e., cognitive and structural social capital. Lower level structures in the island community involve the different forms of social capital producing elements of the cognitive and structural domain. These elements refer to the emergents, either ephemeral or stable in nature. The emergents can exert positive and negative downward causation on the further development of social capital. Moreover, they have purposeful or incidental effects,

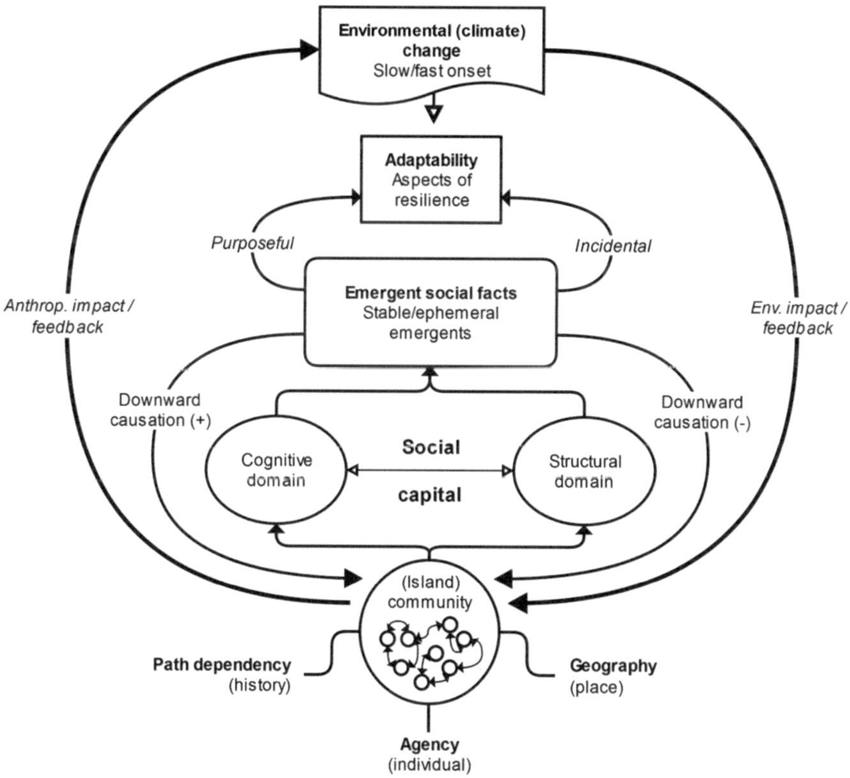

Fig. 6.1 Extended framework of social capital, emergence, and adaptability

which influence adaptability regarding the different aspects of resilience in a context of environmental (climate) change. The dimension of the change is influenced by feedback from the social systems while it also influences the latter directly by promoting and confining processes of downward causation. Each of them has to be understood as a non-linear relationship. Consideration of these factors and reciprocal relationships allows for an understanding of the influence of social capital for community adaptability towards climate change on small islands, generalised here as environmental change on small islands.

While an insular setting cannot be understood as simply implying vulnerability or a specific resilience in a geographic deterministic way, certain developments and modes of societal self-organisation are likely and can be included to better explain island case study analysis. Indeed, social memory and local knowledge can increase adaptability, but not automatically. Likewise, geographical setting, e.g., insularity, does not imply a deterministic resilience. Therefore, all the factors described in this chapter must be understood as complementary, affecting each other, and, accordingly, differing from case to case.

To answer the research question, therefore, it is not only specific factors that shape adaptability towards climate change on small islands, but it is about analysing links, reciprocities, feedbacks and contexts that differ from case to case—implicit in an understanding of complexity. Again, this should stress the rejection of geographic determinism and unidimensional causations as explanatory approaches where either geography or social capital is a single focus of research.

6.3 Theoretical Limitations

The case study of an archipelago and the specific context of climate change adaptation were the basis for the integration of the various theoretical perspectives through in-depth empirical field work. As with any case study research, however, there are implicit theoretical and epistemological limitations, as well as newly emerging perspectives that could not have been included in the research design.

The scope of this book addressed several specific questions. The theoretical framing does include even more issues and the results raised new questions that can be formulated with the theoretical background—not answered with the available data, but nonetheless explicitly acknowledged here.

While the perspective of methodological collectivism allows for the identification and analysis of processes and outcomes of social emergence, it neglects a profound understanding of individual agency. Thus, complementary to a collectivist view, questions about individual motivations and capacities to act arise. Also the role of leadership, regarding individuals but also organisations, needs to be addressed. What are barriers and drivers of emerging leadership?

As mentioned in the theoretical overview on the concept of social capital in Sect. 3.2, social capital does not necessarily imply benefits for all members of a community. Again, here the individual perspective allows for the identification of lack of access to social capital and patterns of exclusion of individuals from a community by means of social capital. Moreover, social capital can promote not only beneficial collective behaviour but also conservatism, which in the context of climate change can imply a lack of adaptability. In addition, it is not always clear whether effects of social capital for climate change adaptation are purposeful or incidental. In fact, the reality behind the constructed distinction changes and formerly incidental social capital can become institutionalised and purposeful.

Despite the insistence upon an alternative view on islands and their relationship with the sea, to study islands on their own terms, the interpretation of the empirical results is influenced by island-mainland thinking. In a context of resilience and social emergence in a particular geographical setting, it is, in fact, difficult to completely neglect this perspective. Therefore, it is important to be aware of it and not to draw speculative and geographically deterministic conclusions. Although the framework and the empirical results address the topic of archipelagos, the importance of inter-island in contrast to intra-island social capital and its consequence for the development of stable and ephemeral emergents is not yet completely clear.

Some of the emergents are not addressed because my analysis started from other assumed effects of social capital for adaptation, such as collective action, information, decision-making, etc. Yet also language, literature and sense of place are of course important and related to social capital as well as emergence. I showed the relationship between the different forms of social capital, emergence and the four aspects of resilience. Here, the role of panarchy needs to be addressed further. How is emergence understood and defined on different levels—local, regional, national and international? Furthermore, (how) can emergence be distinguished on the different levels? Related to the negative effects of social capital from a resilience perspective, path dependence can imply a lock-in or opportunity. When is a certain development promoting adaptability and when is it an obstacle? Again, this is dynamic, non-deterministic and can only be completely understood retrospectively.

References

Berkes, F., & Ross, H. (2013). Community resilience: Toward an integrated approach. *Society & Natural Resources, 26*(1), 5–20.

Manson, S. M. (2001). Simplifying complexity. A review of complexity theory. *Geoforum, 32*(3), 405–414.

Mercer, J., Kelman, I., Alfthan, B., & Kurvits, T. (2012). Ecosystem based adaptation to climate change in caribbean small Island developing states: Integrating local and external knowledge. *Sustainability, 4*(12), 1908–1932.

Ratter, B. M. W. (2013). Surprise and uncertainty—Framing regional Geohazards in the theory of complexity. *Humanities, 2*(1), 1–19.

Sawyer, R. K. (2005). *Social emergence. Societies as complex systems* (Vol. 4). Cambridge: Cambridge University Press.

Urry, J. (2005). The complexity turn. *Theory, Culture & Society, 22*(5), 1–14.

Wilson, G. A. (2015). Community resilience and social memory. *Environmental Values, 24*(2), 227–257.

Chapter 7
Conclusions

Like the Isles of Scilly, many small island communities are confronted with environmental stressors, both now and in the past. These stressors are expected to increase with further climate change. The guiding research question for this work was.

How does social capital help explain collective behaviour as well as opportunities and limitations of societal adaptability towards non-linear climate change on small islands?

Moreover, which implications does an understanding of this have for the theoretical context and for actual adaptation to climate change? To answer the research question, I used a theoretical approach based on complexity theory (social emergence), seen through a lens of social capital and framed by nissology, to understand the behaviour and context of small islands.

7.1 Outcomes

The example of the Isles of Scilly demonstrates that social capital can be important for adaptation to climate change in different ways. Where environmental protection is organised by a very dominant management body, the role of social capital is less relevant to collective action, though it is relevant to disaster response and flow of information; e.g., on St Mary's, where the council and organisations like the Wildlife Trust are very active, social capital is more relevant to disaster response. Trust and a cohesive community encourage people to come together and help each other. On Tresco, however, the estate management is so dominant in terms of environmental and coastal protection that the role of the community is almost negligible. On the other off-islands, where the communities are smaller and more isolated, collective action for the maintenance of infrastructure and sea defences is more important. On islands that have very cooperative communities, as well as skill and machinery, like St Agnes, this is most effective.

J. Petzold, *Social Capital, Resilience and Adaptation on Small Islands*,
Climate Change Management, DOI 10.1007/978-3-319-52225-8_7

My results demonstrate the potential and limitations of social capital for adaptation confirming that a high level of social capital does not guarantee per se higher adaptability. The Isles of Scilly have always been vulnerable to certain risks that are now associated with climate change, such as storm activity, flooding and erosion. This is part of the islands' identity and collective memory. The islanders are accustomed to being self-reliant in times of emergency. These experiences help people to deal with rough weather conditions, transport problems and food shortages, through self-organisation or with the support of local organisations. At the same time, however, the structure of decision-making and regulations has changed in recent decades. The influx of residents from the mainland and the economic shift from agriculture towards tourism has led to a declining number of people from traditional island families. This development leads to a loss of traditional skills of land management but also introduces new modes of living and conducting businesses that can be important for the overall future development of the island communities. Civil society organisations facilitate bringing together local and external resources and mobilising the community.

How relevant social capital is for the adaptation to climate change also depends on the dimension of the vulnerability of the island and its institutional setting. High social cohesion, levels of trust and participation in clubs do not automatically produce high collective action. Social capital has to be mobilised for climate change adaptation; individuals who initiate action and institutions that allow for and promote the integration of community resources influence whether existing social capital can play an important role for adaptation. Leadership, skills and tools shape the effectiveness of social capital for collective action.

7.2 Theoretical Implications

My research results are embedded in the context of non-reductionist social sciences and methodological collectivism.

I used the existing theoretical developments on social capital and applied the theory in a collectivist understanding to the concept of social emergence, contrary to the more common individualist approaches. In this way, my research filled a gap in analytical value of social capital theory for community resilience by showing how social capital is mobilised, and how the mobilisation of social capital is promoted or confined by societal, geographical and historical factors. I reached this understanding by means of interpreting social capital from a perspective of complexity theory. Moreover, my approach contributes to the understanding of social emergence, by explaining the steps between the individual, interaction and emergent social facts. Hence, the connection of social emergence and adaptability can be understood through analysing the role of collective social capital in a context of complex systems.

Every community is different. Therefore, generalised statements about the effectiveness of social capital that neglect such very basic features of a society as its demographic structure and geographic features necessarily lack in explanatory power. A focus on informal collective action and the role of civil society organisations is useful in pointing out the linkages between the micro- and macro-level of society. While the focus here is on collectivism, individual social capital and the role of friends and family networks are nonetheless highly relevant to individual adaptation to climate change and the inner nature of collective action processes. To explain the full significance of social relationships for adaptability, research on individual social capital is valuable, complementary to the analysis of collective social capital and social emergence.

Social capital is multi-dimensional and refers to different levels of a social system illustrated by bonding, networking and linking social capital, as well as cognitive and structural elements. It is not only important for climate change adaptation when used purposefully; also its incidental effects are important. The lack of consideration of the latter in the analysis of social capital has resulted in an incomplete understanding of how social capital is mobilised and becomes relevant to coastal community resilience in response to sea-level rise (cf. Pelling 2011). Incidental and purposeful effects are important for various tasks which are involved in climate change adaptation, including political, economic, collective and individual capacities (cf. Petzold and Ratter 2015).

My results refer to the adaptability of communities in a context of specified resilience towards sea-level rise—not general social-ecological resilience of the system. I identified geographical context, here insularity, as a key factor for understanding the role of social capital for social emergence and how that context influences each element of community resilience. Some considerations should be made concerning the generality of my findings and the validity of my results for small islands in general, larger islands, continental coastal areas or other non-place-based communities.

The results here are based on empirical research on small islands forming an archipelago theorised in a context of place-based communities with a strong relationship to the sea. The insights from the case study describe the potential of social capital under certain conditions such as isolation and remoteness. My results refer primarily to small islands, i.e., islands shaped by common features of scale and location as well as vulnerability to climate change. While most existing literature focuses on SIDS, I here analysed a case study located in the so-called global north and a dependent archipelago. My results showed, that social capital is important not only in economically and politically relatively weak societies. However, I suggest that the features affecting the mobilisation of social capital for a specific purpose (here, adaptation to sea-level rise) are similar, but have a different relevance. This applies especially to the locally specific administrative and economic conditions, influencing the effective mobilisation of social capital. On the other hand, often neglected incidental features of social capital might have similar relevance.

In the geographical sense, the relationship with the sea is especially important because it affects many of the incidental features of social capital that have the potential to increase adaptability. Similarly important aspects are geohazard experience and social memory. Such features, however, could also be extended to non-coastal regions, which face specific environmental changes. On the other hand, the notion of islands as places of transition and ecotones implying a connecting rather than confining character are place-specific and difficult to transfer. The study of islands from a nissological approach allows for the understanding of systems and environments in a context of social emergence. The different dimensions of social capital and multi-scale relevance imply a dynamic definition of systems, which overcomes the conception of physical borders as system boundaries. Thus, islands cannot be considered closed systems. This, however, neither neglects the role of geography nor implies geographic determinism. Instead, the critical study of islands opens up horizons for a non-deterministic research of complexity and geographic place.

7.3 Social Relevance of Results

Linking back to the nissological principles of 'time', 'height' and 'width' (Sect. 3.3. 3; McCall 1994), my research can be socially relevant in different dimensions. In the context of climate change, many small islands have reached turning points and internationally crucial political decisions are being taken ('time'). Thus, my research has to be viewed in a context of possible future trajectories, understanding the role of social capital and which role they can play in such scenarios. Every island faces specific challenges, and results have to be made accessible for local decision-makers and the public ('height/width'). I communicated my results to local stakeholders and inhabitants on the Isles of Scilly for whom the issue of sea-level rise will very likely be of increasing importance. The potential for integrating the local community in future adaptation strategies is high.

In general, however, uncertainties in climate models and the non-linearity in the climate system make it difficult to plan adaptation. Therefore, regional in-depth studies such as this one help in understanding potential local impacts of slow and rapid onset events in a global context, and are important foundations for local and regional policy-making. Understanding limitations and opportunities, obstacles and driving forces regarding social capital has explicit relevance for policy-making and local action initiatives. Consideration of geographical, historical and institutional contexts helps to identify barriers and drivers of adaptability. This can help to remove obstacles as well as to make use of neglected potentials of social capital for adaptation. Thus, my results can be applied to discussions concerning community-based adaptation (cf. Chap. 2) through the integration of local (though not necessarily traditional) knowledge and skills, with expert knowledge, raising awareness and making use of increased flow of information in communities with high social capital. My findings are also relevant to considerations of informal networks and initiatives of collective action in addition to civil society organisations and participatory governance.

Altogether, I avoid transferring results one to one from one island to another and applying a top-down approach. It should be stressed that an approach to resilience and social capital that simply delegates responsibility to communities must be rejected. Islanders should be listened to and empowered to act on their own terms (cf. Betzold 2015). Learning from islands and the combination of traditional and new local knowledge with scientific knowledge can contribute to the understanding of locally implemented adaptation strategies. In climate change discourse, this is an issue that is promoted by island stakeholders (Kelman et al. 2015), but it is certainly generally applicable.

My results show that local issues always have to be considered together with higher scales, e.g., the influence of national and international political and economic conditions, which impact the effective mobilisation of social capital on the local level. However, while climate change is certainly a pressing issue on small islands, the media discourse should not distract from the importance of local circumstances and the fact that local problems are often also related to other causes (cf. Kelman 2014). In a context of analysis of social capital from a resilience perspective, local dynamics become visible and a simplistic perspective on island vulnerability is avoided. In this way, island case studies can provide a basis for a general development of integrative concepts for spreading of knowledge on locally specific adaptability, but also for the dynamic advancement of new local knowledge—a 'new island thinking' (cf. Baldacchino and Clark 2013).

7.4 Outlook

Besides the conclusions and relevance for policy-making, my research results also raised further questions, opening up horizons for further research in the fields of social capital theory, social emergence and nissology.

Further research and combination with quantitative research methods can show how specific social networks and participation in community groups influence the mobilisation of social capital. Features connected to social emergence, such as the role of social memory, language and discourse patterns as stable emergents, need to be addressed in a context of community resilience. Furthermore, how do ephemeral emergents become stable in a context of human-environment interaction? Generally, incidental effects of social capital and emergence need further studying and a firm empirical basis. Similar to the insular features identified on the Isles of Scilly, the question is which role geographic features generally play for incidental effects of social capital without being deterministic.

Climate change research is still struggling to effectively integrate insights from natural and social sciences (cf. Victor 2015). For further integrative research regarding adaptability towards sea-level rise, it is crucial to combine the analysis of small island resilience from a social sciences point of view with locally adapted assessments of the physical environment, including coastal morphology, and detailed assessments of locally distinct patterns of changes in sea level and weather conditions.

The climate change and sustainability discussion will continue as new issues arise and questions emerge about how people deal with their environment. In a context of adaptation to sea-level rise, the perspective on the relationship between people, land and water is insightful. Several questions regarding nissology can be specified for further research. The ecotone approach (Gillis 2014), which focuses on the construction of the margin between land and water, needs further research. If islanders are considered as coastal people, how does that relate to coastal protection? How do people perceive the coast and how do they deal with it accordingly? The concept of ecotones, here, refers to place-specific characteristics in coastal zones, but with a different quality particularly on small islands—places where the intersection of the two ecosystems, marine and terrestrial, is dominantly shaping the environment and potentially also its people. Analyses of this particular feature of ecotones is important for understanding adaptation to sea-level rise, considering the need for more locally sensitive research and planning and regional discrepancies in impacts and adaptability. The concept of ecotones contributes to the understanding of different island topographies, histories and local environmental knowledge that can become relevant for adaptation to risks associated with climate change and sea-level rise. Yet further empirical evidence is needed for a comprehensive theorisation of the ecotone concept in social sciences. Moreover, when is a person an islander? Can this actually be determined? Does it have to be, or does a specific relationship with the coastal environment suffice? People do perceive and identify themselves very strongly as island or mainland people; therefore, it is worth studying critically. Then (how) does a perception or consideration of one being an islander affect their role within community action?

Studying islands on their own terms does not mean neglecting the interconnectedness of islands in the globalised world. As I pointed out, in fact, islands have always been connected to other islands and other continents because they rely on these connections. Therefore, nissological research studying islands as archipelagos —even within the global archipelago (cf. Depraetere 2008a)—can focus on these features of islands: the role of close and distant, regional and island-island relationships, compared to local intra-island relationships. Social capital theory can produce critical insights if integrated into such a context, where small islands can be viewed in a misleading but common conception of being set in the 'ocean as a metaphorical desert' (Rainbird 2007: 4) or rather, as I suggest, 'oases'.

References

Baldacchino, G., & Clark, E. (2013). Guest editorial introduction: Islanding cultural geographies. *Cultural Geographies, 20*(2), 129–134.

Betzold, C. (2015). Adapting to climate change in small island developing states. *Climatic Change, 133*(3), 481–489.

Depraetere, C. (2008). The challenge of Nissology: A global outlook on the world Archipelago. Part I: scene setting the world Archipelago. *Island Studies Journal, 3*(1), 3–16.

Gillis, J. R. (2014). Not continents in miniature: Islands as ecotones. *Island Studies Journal, 9*(1), 155–166.

Kelman, I., Burns, T. R., & Des Johansson, N. M. (2015). Islander innovation: A research and action agenda on local responses to global issues. *Journal of Marine and Island Cultures, 4*(1), 34–41.

Kelman, I. (2014). No change from climate change: Vulnerability and small island developing states. *The Geographical Journal, 180*(2), 120–129.

McCall, G. (1994). Nissology: A proposal for consideration. *Journal of the Pacific Society, 17*(2–3), 1–14.

Pelling, M. (2011). Climate change and social capital. Foresight project. Government office for science. London (International dimensions of climate change, 5).

Rainbird, P. (2007). *The archaeology of islands*. Cambridge: Cambridge University Press. (Topics in contemporary archaeology).

Petzold, J., & Ratter, B. M. W. (2015). Climate change adaptation under a social capital approach —an analytical framework for small islands. *Ocean and Coastal Management, 112*, 36–43.

Victor, D. G. (2015). Climate change: Embed the social sciences in climate policy. *Nature, 520* (7545), 27–29.

Appendix A
Coastal Vulnerability Maps of the Isles of Scilly

(1) Coastal vulnerability of St Mary's: predicted erosion (red line) and extent of still water flooding (blue) in 2105 (Source: Environment Agency 2009c)

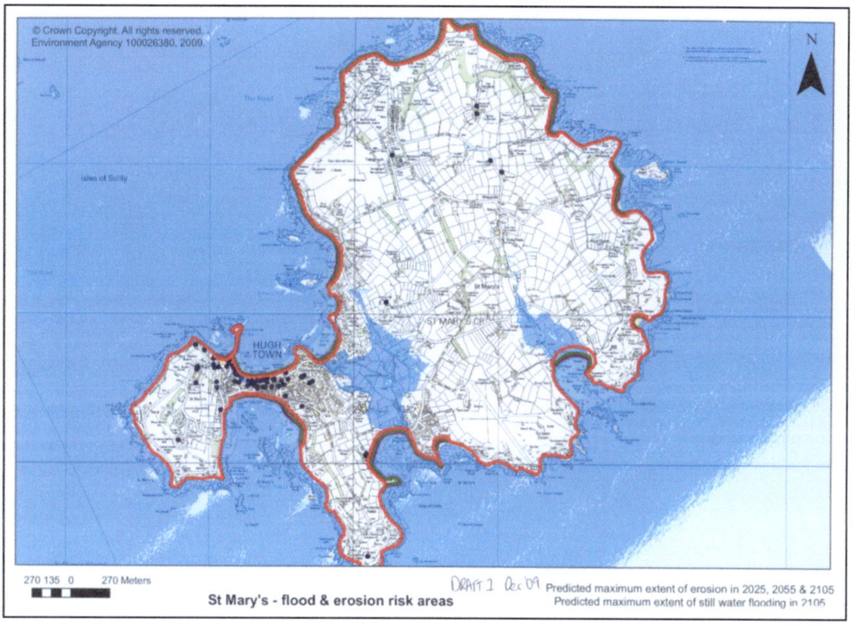

J. Petzold, *Social Capital, Resilience and Adaptation on Small Islands*,
Climate Change Management, DOI 10.1007/978-3-319-52225-8

(2) Coastal vulnerability of Tresco and Bryher: predicted erosion (red line) and extent of still water flooding (blue) in 2105 (Source: Environment Agency 2009d)

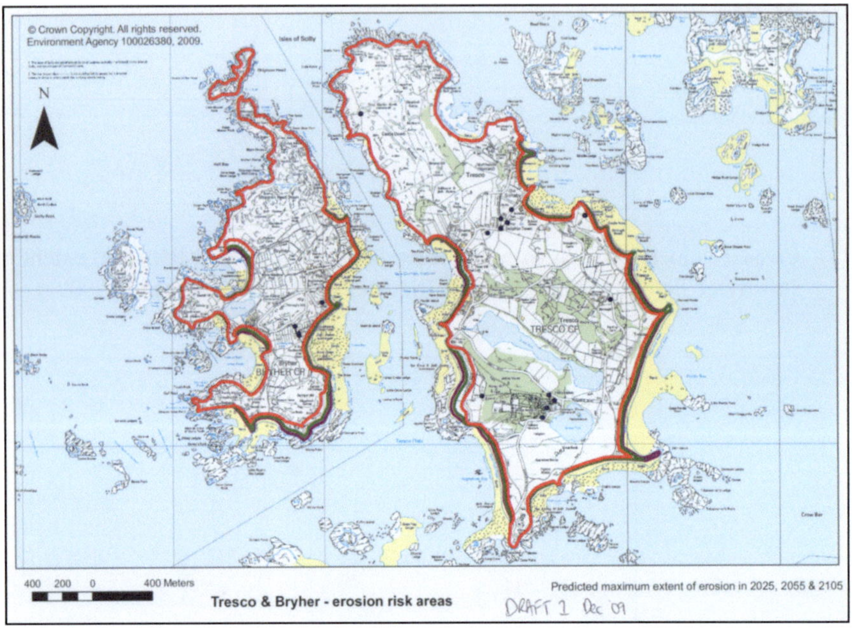

(3) Coastal vulnerability of St Agnes: predicted erosion (red line) and extent of still water flooding (blue) in 2105 (Source: Environment Agency 2009a)

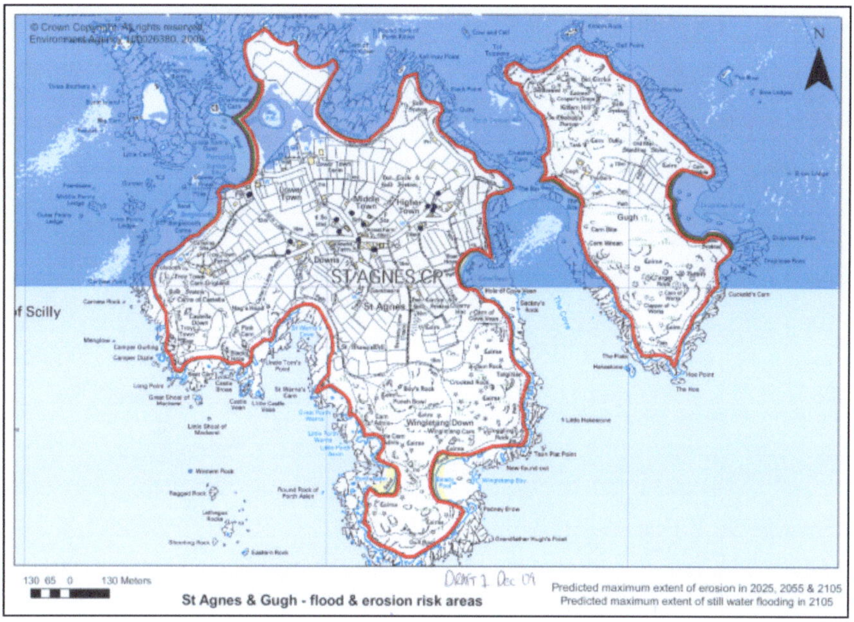

(4) Coastal vulnerability of St Martin's: predicted erosion (red line) and extent of still water flooding (blue) in 2105 (Source: Environment Agency 2009b)

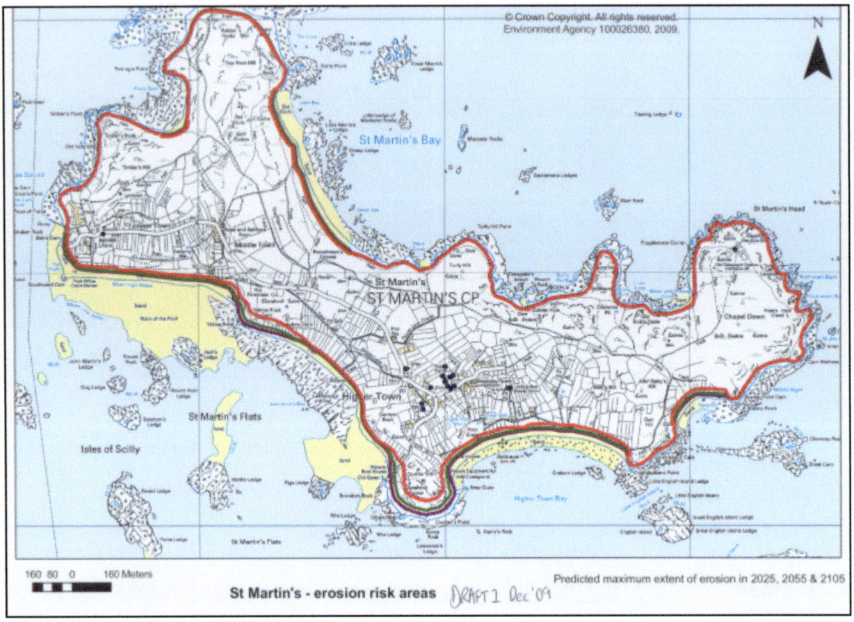

Appendix B
Topographic Maps of the Isles of Scilly

(1) Topographic map of St Mary's (Source: altered, OpenStreetMap 2016d[1])

[1]This and all following maps available under the Open Database Licence: www.openstreetmap.org/copyright.

J. Petzold, *Social Capital, Resilience and Adaptation on Small Islands*,
Climate Change Management, DOI 10.1007/978-3-319-52225-8

(2) Topographic map of Tresco & Bryher (Source: altered, OpenStreetMap 2016c)

(3) Topographic map of St Martin's (Source: OpenStreetMap 2016b)

(4) Topographic map of St Agnes (Source: OpenStreetMap 2016a)

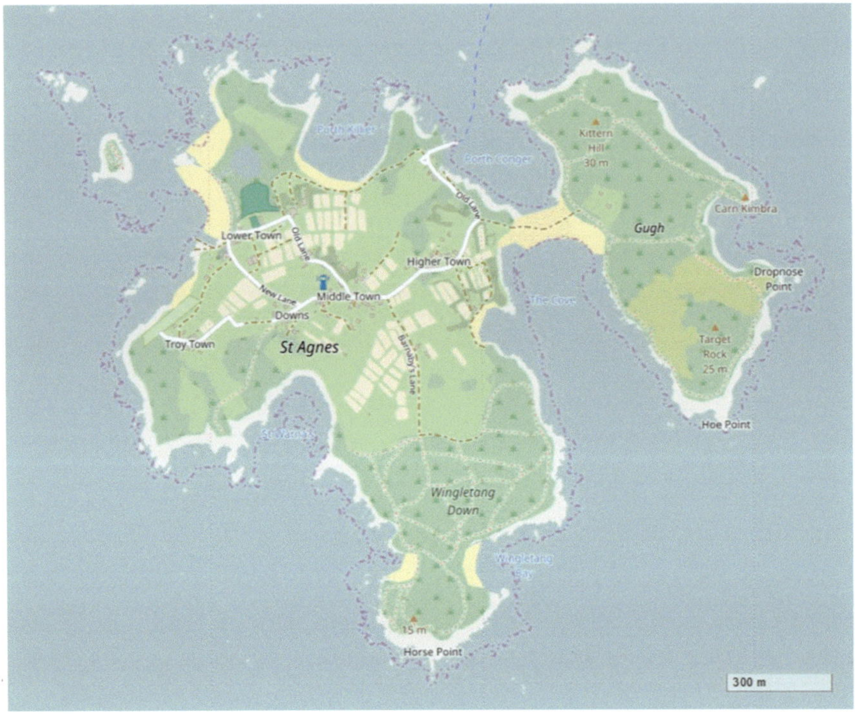

References

Environment Agency. (2009a). St Agnes & Gugh—flood & erosion risk areas. Crown Copyright, Ordnance Survey Licence 100026380.

Environment Agency. (2009b). St Martin's—erosion risk areas. Crown Copyright, Ordnance Survey Licence 100026380.

Environment Agency. (2009c). St Mary's—flood & erosion risk areas. Crown Copyright, Ordnance Survey Licence 100026380.

Environment Agency. (2009d). Tresco & Bryher—erosion risk areas. Crown Copyright, Ordnance Survey Licence 100026380.

OpenStreetMap. (2016a). Map of St Agnes. https://www.openstreetmap.org/#map=15/49.8914/-6.3456. Checked on 19/10/16.

OpenStreetMap. (2016b). Map of St Martin's. https://www.openstreetmap.org/#map=15/49.9649/-6.2859. Checked on 19/10/16.

OpenStreetMap. (2016c). Map of Tresco & Bryher. https://www.openstreetmap.org/#map=14/49.9540/-6.3391. Checked on 19/10/16.

OpenStreetMap. (2016d). St Mary's. https://www.openstreetmap.org/#map=14/49.9209/-6.3104. Checked on 19/10/16.

J. Petzold, *Social Capital, Resilience and Adaptation on Small Islands*,
Climate Change Management, DOI 10.1007/978-3-319-52225-8

MIX
Papier aus verantwortungsvollen Quellen
Paper from responsible sources
FSC® C105338

If you have any concerns about our products,
you can contact us on
ProductSafety@springernature.com

In case Publisher is established outside the EU,
the EU authorized representative is:
Springer Nature Customer Service Center GmbH
Europaplatz 3, 69115 Heidelberg, Germany

Printed by Libri Plureos GmbH
in Hamburg, Germany